Biosimilars

Jean-Louis Prugnaud
Jean-Hugues Trouvin
Editors

Biosimilars

A New Generation of Biologics

Springer

Editors
Jean-Louis Prugnaud
Hôpital Saint-Antoine
Paris
France

Jean-Hugues Trouvin
Université Paris Descartes
Paris
France

Translation from the French language edition 'Les biosimilaires' by Jean-Louis Prugnaud, Jean-Hugues Trouvin (Eds.), © Springer-Verlag France, Paris, 2011; ISBN: 978-2-8178-0036-3
Translated by: Mrs. Laurence Grenier

ISBN 978-2-8178-0335-7 ISBN 978-2-8178-0336-4 (eBook)
DOI 10.1007/978-2-8178-0336-4
Springer Paris Heidelberg New York Dordrecht London

Library of Congress Control Number: 2012949364

Printed on acid-free paper

Springer is part of Springer Science+Business Media (www.springer.com)

Foreword

Biosimilars: A Philosophy?

Setting the Scene for Biosimilars

Biological medicinal products including biotechnology-derived medicinal products (often referred to as "biologicals") have an impressive record in treating numerous serious diseases, and their market is growing faster than that of all pharmaceuticals combined. Insulin produced by using recombinant DNA technology was the first approved therapeutic protein. It entered the US market in October 1982 and subsequently also gained its marketing authorisation in Europe. Since that time, which created, a "gold rush mood" of sorts, the progress in the research and development of these innovative medicinal products has accelerated significantly. At present, several hundreds of biologicals (in the broader sense) have been approved in Europe and the United States, and the number of applications for marketing authorisation is still rising. However, the research and development on biotechnology-derived medicinal products is costly, including also the considerable work one needs to invest into defining and maintaining a well-controlled manufacturing process. Therefore, high costs must often be paid when it comes to treating patients, which represents a burden to health care systems, and thus might limit access of patients to these medicines. The upcoming expiration of patents and/or data protection for the first innovative biotherapeutics has apparently created another "gold rush mood" to work on "generic versions" of products "similar" to the originals and relying in part for their licencing on data from these originator products for their licencing. As we will learn in this book, however, the term "generic" cannot be used for such biologicals that are "copy versions" of licenced products. The most important reason is that biologicals are complex both as regards their structure and their manufacturing process. Even with very sensitive state-of-the-art physicochemical and biological characterisation methods, one cannot conclude that an originator product and a copy version of it are "essentially similar" or even "identical". There could be minute differences, and these differences could have a huge impact on nonclinical and clinical behaviour, such as safety or efficacy. This is why the term "biogeneric" is obsolete, and another term, namely "similar biological medicinal product", or, in widely used jargon, "biosimilars", has been coined. Licencing of biosimilars has already become a reality in Europe, since the European Medicines Agency's scientific Committee for Medicinal Products for Human Use (CHMP) has established a regulatory framework for similar biological medicinal products for their facilitated development, and issued a positive ruling on two "biosimilar" applications in 2006 (Omni trope and Valtropin, both of which are growth hormones). Many aspects of this framework and its application will be discussed in this book. Although the licencing of the first biosimilars was an initial step for other products following the biosimilar route, such an approach also has its limitations in that clinical trials are still required, so no real "generic" route is yet possible. Some general aspects are, however, already clear now: (1) guidelines for the development of biosimilar products can be implemented, (2) many aspects from "comparability exercise" that regulators are already well accustomed with from changes in manufacturing processes of biologicals are applicable and have been readily applied already, (3) as shown in the case of Valtropin, the use of different host cells for the biosimilar product and the comparator in principle can be possible, and (4) although required, the clinical programme might be abridged as compared to a full development and has the primary aim of establishing "similarity". The main principles for the development of a biosimilar medicinal product in the European Union currently include:

- Demonstration of "biosimilarity" in terms of quality, safety and efficacy to a reference product that is licenced in the EU. Therefore, all main studies should be strictly comparative, and the same reference product should be used throughout the development programme.
- Use of a 'sensitive' test model (and here patients are also included, as discussed further below), able to detect potential differences between the biosimilar and the reference product.
- Demonstration of equivalent rather than non-inferior efficacy, as the latter does not exclude the possibility of superior efficacy (only an "equivalence" trial can normally *a priori*, on formal grounds, establish "similarity").

This last point is a frequently asked question: What if the biosimilar is indeed more efficacious than the reference medicinal product? Would that not be desirable for the patient? However, the answer is clear: a superior efficacy would not only contradict the assumption of similarity, and thus potentially preclude extrapolation to other indications of the reference product, particularly those with different dose requirements; it could also imply safety concerns, since in case of higher potency the product could also be more "efficacious" in an unwanted sense, and safety issues could arise when using the dose(s) recommended for the reference product, especially for drugs with a narrow therapeutic index. If similarity with the reference product has been convincingly demonstrated in a key indication, extrapolation of efficacy and safety data to other indication(s) of the reference product, not studied during development, may be possible if scientifically justified. This is sometimes not straight forward, since it usually requires several prerequisites like involvement of the same mechanism(s) of action for each of the indications, or for example the involvement of the same receptor(s) for mediating this mechanism(s) of action. This is one of the aspects that will also be discussed later in this book. It is clear that, although a reduction in the data package is possible for a biosimilar, the pre-licencing data package is still substantial. The dossier for the CMC part (Chemistry, Manufacturing, Controls, or also the "quality" dossier in regulatory jargon) has even higher requirements than a "stand-alone" development of a new biological drug: First, a full quality dossier has to be provided that meets regulatory standards to the same level as a novel biological medicinal product. Second, on top, a comprehensive comparability exercise to the reference product on the drug substance and drug product level is required, which is an additional requirement, exclusive to a biosimilars. Likewise, human efficacy and safety data are always required, but—and this is at the heart of a biosimilar development—to a lesser extent than for the development of a novel medicinal product. Here, and in the possibility to omit certain non-clinical studies usually required for a novel compound, lies the possibility to reduce the costs of development considerably. The amount of possible reduction in non-clinical and clinical data requirement depends on various considerations, including how well the molecule can be characterised by state-of-the-art analytical methods, on observed differences between the biosimilar and the reference product, and on the clinical experience gained with the reference product and/or the substance class in general.

Biosimilarity: A Philosophy?

The difficulty for the developers of any biosimilar is that there is usually no direct access to originator companies' proprietary data. The developer of a biosimilar, thus, has to purchase the reference medicine from a pharmacy and then purify the drug substance, and engineer a process to produce the biosimilar—that is, the development of a biosimilar requires the establishment of a new manufacturing process "from scratch". If the European biosimilar framework had required an identical manufacturing process, then this would have automatically made biosimilar development programmes difficult if not impossible. It is acknowledged in the respective biosimilar guide-line (*Guideline on Similar Biological Medicinal Products Containing Biotechnology-Derived Proteins as Active Substance: Quality Issues CHMP/49348/05*) that "*it is not expected that the quality attributes in the similar biological and reference medicinal products will be identical*". This follows from the principle that a biotechnological product is defined by the way it is manufactured, including all process-related and product related impurities, microheterogeneities, excipients, etc ("process determines product", or "the process is the product"). Due to the complex production method of biological medicines, the active substance may differ slightly between the biological reference and the biosimilar medicine—what "slightly" will mean, however, will unfortunately be a case-by-case decision based on data and involving various considerations like the complexity of the molecule in question, the inherent variability known from the reference medicinal product, the potential clinical impact etc. One can easily recognise that a biological is therefore more than just the active substance, and includes the aforementioned impurities etc. "Slight" differences can have a major impact, but, theoretically, conversely some differences e.g. in impurities may have no impact at all. So how to solve this problem to establish biosimilarity? First of all, there must be a backbone of extensive and state-of-the-art data from chemistry, manufacturing and control ("quality data") that not only satisfy pharmacopoeias, but is also strictly comparative to the reference product. This serves as the basis for abridging the non-clinical and clinical data requirements. When it comes to clinical data, there is at least one decisive difference to clinical development programmes for "stand-alone" developments like biologicals with a novel mechanism of action, and this difference in concept is sometimes hard for clinicians to swallow: The aim of a biosimilarity development programme is not to establish benefits for the patient—this has been established already by the reference product years ago. The aim of a biosimilarity development programme is to establish biosimilarity and if there are clinically relevant differences, by employing a clinically relevant model. Indeed, patients are seen here, formally speaking, as a "model" to establish biosimilarity. This means that the trial design including the primary endpoint, secondary endpoints, choice of patients etc may follow a different philosophy than for a novel compound. For example, for a clinical indication that can present itself in different severities, it may be unwise to include patients suffering from different grades of the disease. If there is—despite randomization and/or stratification—an uneven distribution of these numerous factors like different disease history, different pre-treatments, different disease presentation at study entry etc, then differences between the two study arms comparing the reference biological medicinal product with the biosimilar may be difficult to interpret: Is a perceived difference related to differences in the molecules? If so, what if there are no measurable differences on an analytical level? Or could differences be explained by differences in the patients between the groups? In other words—by not focusing on a homogeneous patient population, it could well be that the endpoint measures differences in disease manifestation and not differences between the molecules. Likewise, following this philosophy, the most sensitive clinical end-point may be more suitable than an endpoint establishing clinical benefit. Currently, anticancer biologicals utilising a cytotoxic mechanism of action are not yet approved as biosimilars; however, upcoming discussions will have to elucidate on what endpoint to choose for such scenarios—should it be a more sensitive and measurable end-point, e.g. tumour response rate, or should it be a more clinically relevant endpoint like overall survival of cancer patients? The tumour response rate only measures the action of a drug, not patients' benefit. A highly active compound that yields in a high tumour response rate could, at the same time, be considerably toxic, and thus reduce survival, which would obviously not be a benefit for patients whilst still meeting the endpoint "tumour response rate". Therefore, for new compounds, a time-related benefit endpoint is usually required, e.g. overall survival. However, one could argue that the survival benefit has already been established years before by the originator product, and that for the biosimilar this does not have to be repeated. Such considerations are currently hotly debated. Now that biosimilars have become a reality, developers are extending their reach to more complex molecules—including monoclonal antibodies that are much more complex than currently licenced biosimilars (e.g., growth hormones). It is, therefore, time to discuss the current state-of-the-art, condensing all relevant aspects within a single book. I am sure that this book will not only be useful for developers of biosimilars or for regulators—I do think that also physicians, who are the "users" of biosimilars, should know about how biosimilars are designed and developed, since they are clearly different from generics, which surely has clinical implications.

Christian K. Schneider
CHMP Working Party on Similar Biological (Biosimilar) Medicinal Products Working Party (BMWP), European Medicines Agency, London, United Kingdom

Paul-Ehrlich-Institut,
Federal Agency for Sera and Vaccines, Langen, Germany

Twincore Centre for Experimental and Clinical Infection Research

Preface

The biological medicinal products' market has considerably expanded since 1998. The worldwide sales for these types of medicines have increased much faster than other types of medicines—12 % on average between 1998 and 2007 versus only 4 % for the sector beside biological medicinal products. The share of these in the global market will have risen from 10 to 15 % between 2007 and 2012, according to IMS (*Intercontinental Marketing*[1] lists 633 biological medicinal products are being developed worldwide to treat more than 100 diseases; including: 254 drugs developed for cancer; 162 for infectious diseases; 59 for auto-immune diseases and 34 related to HIV/Aids pathologies).

Numerous recombinant proteins are currently in the public domain after expiration of the patents that protected them, thus they are an interesting target for classic generics companies. If the pressure put on by institutions that provide payment services and a simpler licencing process have contributed to their very large development, then the difficulty to develop copies of biotechnological products could be a factor of weaker progression.

The term "generic medicinal product" is used to describe a medicine that has an active substance made of a small, chemically synthesised molecule with a well-known structure and a therapeutic action equivalent to the original product's. Generally, the demonstration of bioequivalence with a comparator through bio-availability studies is enough to deduct the therapeutic equivalence between the generic and reference medicine. This approach is not considered sufficient for the development, evaluation and approval of a biological medicine claiming its similarity to a reference medicine because of the molecular complexity and the difficulty to characterise active structures. On top of that, efficacy and therapeutic safety may be influenced by the biological source and the manufacturing process. Clinical studies are therefore necessary to demonstrate the efficacy and safety of these copies. As these copies are not identical to their originator but only "similar", they are called "biosimilars", as a contraction of the official European designation of "biological medicinal product similar to a reference biological medicinal product." Other designations can be found in the literature as "biogenerics" but this term can not be retained because of the "only similar" feature that

[1] IMS Health analyse Dévelloppement and Conseil, juillet 1998.

a copy may have. In the U.S.A., the term *follow-on biological product* (FOBP) is used to designate the copies of bio medicinal products. The World Health Organisation (WHO) uses the term *similar biotherapeutic product* (SBP) to designate biosimilars.[2]

The purpose of this book is to show how biosimilars are developed, what the criteria and aspects that are taken into account for their licencing are, how patients safety is preserved, what it is about the particular angle of immunogenicity, what response must be considered concerning substitution and interchangeability of these products, what particular follow-up must be implemented (in terms of pharmacovigilance and traceability) and what the perceptions of the players, prescribers and dispensers of these products are. Biosimilars are medicines destined to be present in doctors' therapeutic tool box. Then, this book tackles the certain aspects of strategies underlying the use of biosimilars and the resulting medical responsibility, if this latter may ever be particular for this new type of medicine.

This book limits itself to the analysis of European licencing of biosimilars as it is currently on a worldwide basis, the most advanced regulation since the first 2001 guidelines that allow with time to build constructive supports designed to help industrialists or companies to develop biosimilars.

The marketing of biosimilars is characterised by many more barriers than for generics; that is to say, developments necessary in order to master their manufacturing and their quality, safety and efficacy evaluation. The biosimilars market has induced high development costs, and higher risks; it needs a longer development time and an expertise related to the clinical development of these products. Biosimilars development strategies are not the same as those of generics. Their development complexity as well as their production costs will favour companies with significant financial resources, experience in the field of biological medicinal products production and even an expertise in marketing innovating products.

In relation to biological medicinal products development costs, the average cost of treatment per patient for this category of products is much higher. Worldwide, 7 medicines of the "top 10" sales will be in 2014 products of biological origin and their individual cost will comprise between 10,000 and 100,000 Euros per person. The appeal of the marketing of copies similar to the original product ensuring the same level of quality, safety and efficacy is obvious for the organisations that provide payment services, as well the as the patients with incomplete or no coverage. However, because of the complexity of the development and production of biosimilars, a reflection on the necessary reduction of these products' costs is needed. Will this reduction be as significant as for generics?

To the effect of the cost less attractive for biosimilar prescriptions may be added a stronger reluctance to use biological products (in their vast majority) that only

[2] Medicines in development. Biotechnology. Billy Tauzin. 2008Guidelines on Evaluation of Similar Biotherapeutic Products (SBPs). WHO/BS/09.2110. Source: EP Vantage June 2009; Evaluate Pharma: World preview 2014, report.

specialists could master. The statute of prescription and dispensation of biological medicinal products is subject to international regulations based upon international studies that often underline the complexity of their use and of patient follow-up. Biosimilars do not escape from these rules and obligations. Consequently, it is important for the doctors who are in charge of treating patients with biological medicinal products to know the respective contributions of each available medicine, whether it is a reference product or biosimilar. This book is aiming at simplifying the approach of products as complex as biosimilars. To do so, it reminds the reader:

- the complexity of biotechnology products and their mode of production;
- factors of safety for their approval application and their marketing authorisations;
- the risk analysis and possibilities of interchangeability;
- the French authorities' interdiction of substitution of a prescription by the pharmacist;
- the analysis of the rules laid down by some learned societies or professional associations, for a better follow-up and a better prescription of biosimilars.

Today, thanks to a centralised and specialised European licencing process, the number of biosimilars that got their marketing authorisation is limited but advanced when compared to some countries, like the United States or Japan. In some other countries, some companies offer copies of biotechnological products without having to apply for approval based on an approach equivalent to that of Europe. Currently, Europe is essentially concerned by the growth hormone haematopoietic growth factor G-CSF (*Granulocyte Colony Stimulating Factor*) and erythropoietin. The next approval applications should be affected by the arrival of copies of other therapeutic proteins like insulin, interferon α and monoclonal antibodies. Similarly, copies of medicines other than therapeutic proteins should reach the biosimilars market, such as fragmented heparins; for which a recommendation by the licencing authority has been published. This sector of medicine is thriving. It requires a professional's deep knowledge in order to maintain the quality and the safety of patient treatments.

Jean-Louis Prugnaud

Contents

Contributors

Christos Chouaïd Service de pneumologie Hôpital Saint-Antoine, UMR Inserm S-707, 184 rue du Faubourg Saint-Antoine, 75571, Paris Cedex 12, France

Kowid Ho Département de l'évaluation des médicaments et produits biologiques, Agence Française de sécurité sanitaire des produits de santé, 143 Boulevard Anatole, 93200 Saint-Denis, France

Didier Kamioner Service de cancérologie et d'hématologie, Hôpital Privé de l'Ouest Parisien, 14 avenue Castiglione del Lago, 78190 Trappes, France

Francis Megerlin MCF droit et économie de la santé Liraes, Université Paris Descartes, 4, avenue de l'Observatoire, 75006 Paris, France

Mira Pavlovic DEMESP Haute Autorité de santé, 2, Avenue du Stade de, Saint-Denis La Plaine Çedex, 93218 France

Jean-Louis Prugnaud Département de l'évaluation des médicaments et produits biologiques, Agence Française de sécurité sanitaire des produits de santé, Président Commission thérapie cellulaire et thérapie génique, 13 avenue Jean Aicard, 75011 Paris, France

Jean-Hugues Trouvin Département des sciences pharmaceutiques et biologiques, Université Paris Descartes, Avenue de l'Observatoire 4, 75006 Paris, France

Biologicals' Characteristics

K. Ho and J.-H. Trouvin

Introduction: From Generics to Biosimilars

In the field of chemical medicines (active substances derived from chemical synthesis), the "generic" procedure is well known and can be used once the protection period has expired (patents and Marketing Authorisation). A "generic drug" is a medicine with the same qualitative and quantitative composition in active substance(s), the same pharmaceutical form as the reference drug, and whose bioequivalence with the reference medicinal product has been demonstrated by appropriate studies. The Marketing Authorisation application file for a generic drug is, compared to a new medicinal product, much more simplified insofar as the usually required non clinical and clinical data are reduced to a comparative bioequivalence study versus the "reference" medicinal product.

Today, in the field of biological drugs, and more specifically of so-called "biotechnological" drugs (see infra, definitions), patents and other data protection are falling in the public domain (ex: Insulin, Somatropin, Erythropoietin, etc.). Mirroring the generic approach implemented for chemical medicines more than thirty years ago, the question arises as to open the same possible development of "copies" of these biological/biotechnological drugs and to have these copied drugs approved, following the same simplified "generic" procedure.

K. Ho
Département de l'évaluation des médicaments et produits biologiques, Agence Française de sécurité sanitaire des produits de santé, Boulevard Anatole 143, 93200, Saint-Denis, France

J.-H. Trouvin (✉)
Département des sciences pharmaceutiques et biologiques, Université Paris Descartes, Avenue de l'Observatoire 4, 75006, Paris, France
e-mail: jean-hugues.trouvin@parisdescartes.fr

J.-L. Prugnaud and J.-H. Trouvin (eds.), *Biosimilars*,
DOI: 10.1007/978-2-8178-0336-4_1, © Springer-Verlag France 2013

For scientific and technical reasons that will be detailed hereafter, the simplified approval application procedure—applicable to chemical generics—has not been deemed fit for biological and biotechnological drugs and an alternative route has been developed. The specific strategy, known as "biosimilar", for the development, assessment and licensing of those drugs claimed "similar" to a reference biological product has been put forward to take into account these products' specificities and to ensure that the benefit/risk ratio of copies of reference biological products has been correctly evaluated before they are put on the market.

Definitions

Biologics

A definition of a biological medicinal product is given in the European Directive 2001/83/CE (Annex I). It meets two characteristics that justify the regulatory requirements and criteria applied to them when they are evaluated for the marketing authorisation application: "A biological medicinal product is a product that has a biological substance as an active substance. A biological substance is a substance produced or extracted from a biological source and that needs for its characterisation and the determination of its quality a combination of physico-chemical-biological testing, together with knowledge of the production process and its control."

Included in biological medicinal products are vaccines, drugs derived from human blood and plasma, as well as any substance extracted from animal or human fluids or tissues, but also products known as "biotechnology" (see infra), and more recently, innovating therapy medicinal products; whose definition is given in the European Regulation 1394/2007.

The classification as a biological medicinal product leads to, for regulatory purposes, the setting up of more demanding evaluation criteria; due to these products' complexity and their production processes that make their final quality more difficult to guarantee and master.

Genetically Engineered Products

Among active substances of biological origin is found a particular class of products known as "genetically engineered", as detailed in European regulation 2309/93, annex I, part A: Medicinal products developed by means of one of the following biotechnological processes:

- recombinant DNA technology;
- controlled expression of genes coding for biologically active proteins in prokaryotes and eukaryotes, including transformed mammalian cells;
- hybridoma and monoclonal antibody methods.

In the context of biosimilars, genetically engineered products are mainly represented by proteins known as "recombinant proteins", because they are expressed and produced by biological systems (bacteria, yeast, human or animal cells, insect or plant cells, transgenic plants or animals) that have been genetically modified, by inserting specific genetic sequences coding for a therapeutically interesting protein (see infra, production process).

The most famous examples of "recombinant proteins" used in therapeutics are insulin, growth hormone, erythropoietin, hematopoietic growth factors and, more recently, monoclonal antibodies. It is on these recombinant proteins that the first "biosimilars" approach has been designed.

Biological Products Complexity and Examples

The definitions of biological and biotechnological medicinal products, detailed above, permit the identification of these products' main characteristics that underlie the interrogations and technical difficulties routinely met to ensure a production as well as a consistent and reproducible quality control of these products. These characteristics also explain why the simplified approach of generic drugs is not strictly applicable to "copies" of biologicals.

A biological substance is inherently a complex molecular structure. This complex structure is difficult, even impossible, to get by chemical synthesis. So, whereas it is possible to chemically synthesize a twenty- or so amino-acid peptide, it is impossible to produce by chemical synthesis a protein as complex as coagulation factor VIII, growth hormone, or even insulin, though this one is a relatively simple polypeptide.

This molecular complexity also explains the recourse to biological sources in order to extract or produce biological substances. We'll see below all structural elements to consider for the characterisation, production, and quality control of a biological substance before qualifying it for therapeutic use and establishing acceptance criteria for each batch of medicinal product.

The second difficulty that springs from the complex structure of the molecule of interest resides in the technical analytical means to study in detail these various structural aspects of the molecule of interest (or of the molecular population), before considering an administration to the patient. As already stated, synthetic chemistry cannot produce these complex molecules. Consequently, in order to produce these so-called complex molecules, "biological" production systems—with their stream of variability and complexity—will be put in place. Let's remember that, in the definition of a biological medicinal product, the producing process is an integral part of the product ("as well as knowledge of its manufacturing process"). In the field of biological medicinal products it is commonly said that "the process makes the product" or the process takes part in the product definition.

We will successively consider the three elements of complexity, stated above, to illustrate what technical points have to be mastered before ensuring that a biological product, resulting from production system A, is identical (or declared identical) to a same biological product, this one resulting from production system B.

These complex elements will be studied while using the elements that condition the "quality" global profile of the final product in order to illustrate with precise examples a typical quality profile of a recombinant proteins. It is that "quality" profile which also conditions the efficacy and tolerance profile of the medicinal product administered to patients.

Protein Molecular Complexity

In living matter, proteins are structures essential to the organisation of any organelle or organism; they play a central role in terms of structure (for instance, membrane proteins), of metabolic activity (enzyme, cytokine, hormone), or immunologic (immunoglobulins). This protein meets structural characteristics that condition its biological activity, the duration of its activity and presence in the organism (the term half-life is used), and at last its capacity to be recognized by the organism as a known structure (the self) or unknown (the non self) and trigger or not from the receiving organism a defence reaction (neo-antigenicity risk that will be discussed later).

Concept of Primary, Secondary, Tertiary and Quaternary Protein Structure

First of all, a protein defines itself by its primary structure (made of a chain of amino acids) in a determined order (the protein sequence is under the sway of determinism of the gene coding for that protein; mastering the gene coding for the protein of interest will be mentioned in the production process).

The correct sequence of amino acids will determine the molecular constraints, with the consequence of forcing the protein to organise itself in a defined spatial structure. The organisation is described in beta sheets and alpha helices that can be repeated and alternate all along the amino acids sequence (Fig. 1).

This spatial organization can be geopardized, depending on the physico-chemical surroundings to which the protein will be successively exposed, during the step of expression and production in the cell medium, and then during the purification process (see infra, description of the production process) that may lead to pH or denaturing oxido-reduction conditions. Not to mention the various final formulation steps (here we speak of the medicine preparation in its final form as an injectable or freeze-dried preparation) that may lead to instability, cleavage, or denaturing reactions.

Thus the intrinsic protein structure cannot be reduced to a sequence of amino acids alone. Its spatial folding and the preservation of this folding must also be taken into account, as well as the spatial, possibly multi-meric organisation during the medicine's all-life duration.

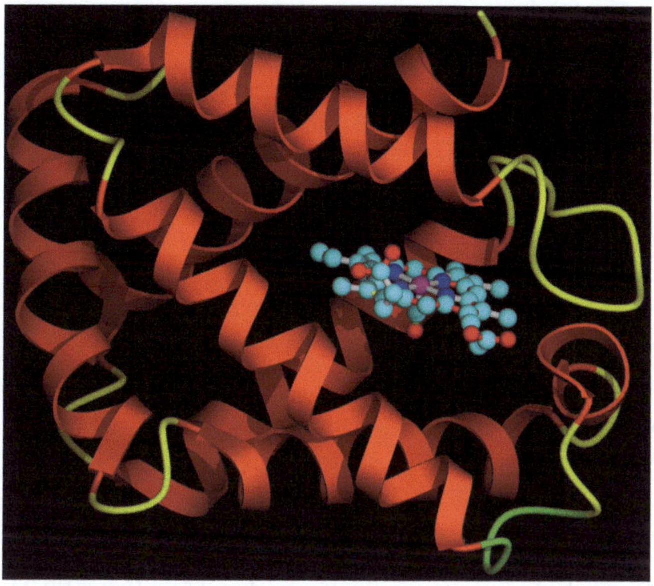

Fig. 1 Tri-dimensional rendering of a protein, with its conformational folding necessary for its activity. Let's notice how complex this protein structure is, compared to the chemical molecule (in *blue*) on which the protein is active. Hemoglobin molecule on *black background* © Pawel Szczesny # 7041692

Figure 2 schematically summarises the four levels of structural organisation of a protein molecule. This conservation of spatial structure, conforming to the "natural mode" when it exists, is essential on one hand for maintaining the protein biological activity, and consequently for its therapeutic activity (for instance, a partially denatured enzyme will not develop its activity, as it is the case of a coagulation factor); and on the other hand, to ensure that the protein will not be recognized as "foreign" by the receiving organism. Indeed, a protein presenting one or several non-conformed parts of the molecule could induce a reaction of defence (immunogenic reaction) within the organism with formation of antibodies directed against the protein of interest.

Concept of Aggregates and Risks of Degradation

The sometimes complex spatial organisation of protein molecules exposes to another risk of the protein non-conformity, that of aggregate formation.

The consequences of protein molecules aggregating between them are multiple and relatively difficult to predict and, sometimes, to detect. Their impact is, however, always negative for biological activity, as well as for clinical tolerance and immunogenic risk.

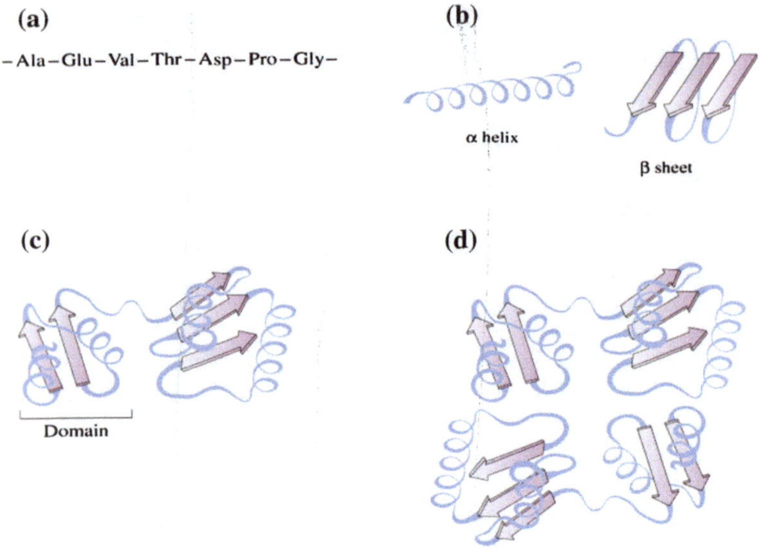

Fig. 2 The four levels of protein organisation (figure adapted from Horton HR, Moran LA, et al. (2002) Principles of biochemistry, 3rd edition). Primary sequence, defined by the gene coding the protein, gives its sequencing to amino acids building the "polypeptide chain." Secondary structure explains the polypeptide chain folding in space. Two types of structures may be noticed, one in alpha helix, the other in beta-sheet. Figure 1 shows an example of protein demonstrating sheets and helix alternating. The organisation in sheets and helix depends on the amino acid sequence (therefore on primary sequence), function of electrostatic forces and degrees of rigidity imposed by each of the amino acids. Tertiary and quaternary structures are adopted by proteins function of their microenvironment and own chemistry. All proteins don't adopt and don't necessarily need to be organised in quaternary structure. **a** Primary structure. **b** Secondary structure. **c** Tertiary structure. **d** Quaternary structure

Local environmental conditions (pH, ionic force, osmolarity, etc.) to be met by the protein all along the production process, followed by purification, and finally of pharmaceutical formulation are crucial. They will induce or not the aggregating behaviour or other degradation mechanisms. Thus, along the development of this production/purification process, but also during the pharmaceutical formulation of the finished product (form under which the product will be "packaged" to be administered), physico-chemical factors potentially inducing such a behaviour of aggregating or instability will be studied in order to suggest systems and/or pharmaceutical formulations that limit these risks and ensure a stable quality from one batch to the other, and stability in the period of claimed shelf-life.

Concept of Post-Translational Modifications: Glycosylation Example

A protein is not characterised only by its primary structure (correct sequence of amino acids) or its spatial conformation (secondary to quaternary structure, Fig. 2);

most often it has additional characteristics acquired during the cellular process of protein synthesis. These are called "post-translational modifications," due to the fact that they occur once the gene (nucleic acids sequence) has been translated into the corresponding protein sequence (the amino acid chain). These modifications are also designated as "maturation phase" essential before the release/secretion of cell proteins. These modifications consist of the grafting on defined amino acids of one or several chemical/biological groups; as, for instance, phosphate or sulphate groups, or sugars (then it is called glycosylation) that modify the global charge and physico-chemical or biological characteristics of these "mature" proteins and condition what they'll become in the organism. It is important to remember that these post-translational modifications are occurring on specific sites of the protein (see below) are not controlled by the gene that expresses the protein sequence, they are instead specific to each cellular kind (notably function of the enzymatic equipment of the cell line expressing the protein of interest and the cell culture conditions (see infra, production process)). These sometimes complex chemical reactions are thus not controllable by of mastering the gene sequence, but by mastering of production conditions in which the cell line chosen to express the recombinant protein will be put. Let's notice that these maturation reactions are absent inside prokaryotic organisms (bacteria), or very simple inside inferior eukaryotes such as yeasts. This notably explains that depending on glycosylation characteristics of the protein of interest, only a "mammalian" cellular system could be considered for production. To give an example, let's detail the glycosylation reaction and its consequences upon characteristics and reproducibility of recombinant proteins and clarify the possible consequences brought by a change of production system, (a change that may occur at all levels, from the nature of the producing cell, to the conditions of culture, and finally of purification). Glycosylation is the most frequent post-translational modification. The chemical modifications introduced are very complex due to the glycanic structures that are added to the protein skeleton. It is that complexity that adds to the global complexity and variability of a mature protein. In a few words, let's remind you what the protein glycosylation step consists of in endoplasmic reticulum and Golgi apparatuses; specific structures of a cellular organisation. A glycosylation consists of branching on the protein, on determined amino acids (for instance, for N-glycosylation, Asn which is in the Asn-X-Thr sequence), sugar groups such as mannose, fructose or galactose following a well-determined orders (Fig. 3). These glycosylation chemical reactions will lead to the making of "sugar chains," more or less complex and diversified, considering all the possible attaching combinations (number of antenna(e) on a glycosylation site, and the nature of sugars making up this antenna), even if some mandatory sequences are found in each structure.

Finally, the end of the sugar chain is most often capped by a sialic acid in the form of neuraminic N-acetyl acid (NANA) in Human cells, when for many mammals a part of the sialic acid is in the form of neuraminic N-glycolyl acid (NGNA) because the gene which codes for the enzyme that allows the NANA form to become NGNA, is muted and inactive in humans. This species specificity is to be taken into account when has to be chosen the cellular system of

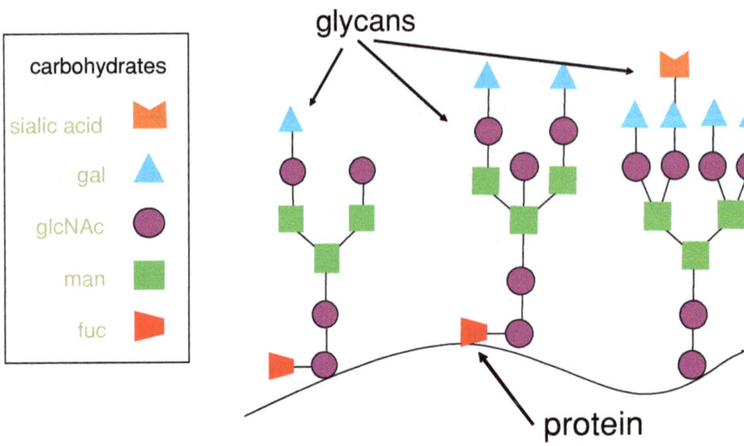

Fig. 3 Schematic drawing of carbohydrate residues (or glycanic structures) present on some protein sequences. Glycanic structures are obtained by combining the sugar group's nature Gal = Galactose, Man = Mannose, Fuc = Fucose, glcNAc = N-acetyl glucosamine and its organisation in antennae (mono, bi, even tri-antennae). Let's also note the presence of a "sialic acid" group that sometimes caps antennae's ends. The sialic acid groups are notably contributing to the protein molecule's half-life

expression/production of the recombinant protein of interest, to ensure that the sialylation is as close as possible to the human form. The mature protein, so "glycolysed" and more or less "sialylated," gets some characteristics that are more or less acidic with a changed isoelectric point (pI). Consequently, at the end of post-translational modifications (maturation) and considering here the glycosylation step alone, the mature protein will show itself not as a single molecular form, but as a mix; a molecular population with the same basic protein structure (primary sequence imposed by gene sequence) on which various types of sugar chains will have been attached, giving then to each protein molecule its own pI. The analysis of this molecular population evidences not a single "form", but a series of isoforms on which a qualitative and quantitative study can be performed by appropriate analytical techniques that separate the various isoforms; for instance, according to their charge.

Generating different isoforms, of which relative proportions are functions of the cell culture medium conditions, induces the notion of a protein "microheterogeneity." So, thanks to these "post-translational modifications" specific to the protein of interest (but also to the expression/production system, as well as the purification scheme), a protein is characteristic because of its "glycosylation profile;" described most often by a series of visible and quantifiable bandwidths, by separation methods of isoelectrofocalisation type. Preserving the glycosylation profile will guarantee the biological activity and the tolerance profile in patients.

Post-translational modifications, usually illustrated by the glycosylation profile, are intrinsic quality criteria of the protein as well as critical parameters to consider during the assessment of the production process and its reproducibility, notably when changes are introduced in the production method, and *a fortiori* when a new manufacturer offers a "biosimilar" version of a reference protein.

Indeed, for the new producer of a given glycosylated protein, one could fear an isoform distribution different from that of the original molecule. This different isoelectric profile—sometimes hard to distinguish by the only analytical methods offered by the manufacturer, will potentially have an impact on the pharmacokinetics or the biological activity of the therapeutic protein. Then it will be the pharmacological and/or clinical data that will reveal the sometimes subtle change in isoform distribution when the quality control analytical data are detecting no noticeable difference.

Although some studies suggest that the consequence of a different isoelectric profile mostly concerns the neo-antigenicity risk, it seems that this phenomenon rather impacts the half-life of the molecule which will be more or less rapidly eliminated by the receiving patient body. Indeed, the sugar chains, notably depending on their sialic acid capping, protect the protein from capture and degradation by hepatic cells.

Thus a recombinant protein will have to have an adapted glycosylation, as well as a correct sialic acid level (in the NANA form), to not to be eliminated too quickly and keep a sufficient pharmacological activity and reduce any potential to generate in patients a defence reaction with formation of antibodies to the protein of interest.

Other Modifications Linked to the Process and/or Conservation/ Formulation

In the paragraphs above, we have mentioned the different critical parameters of a medicinal protein molecular structure. These functional attributes (primary and spatial structures, state of non-aggregation, glycosylation profile and other post-translational chemical modifications) are to be considered along the whole chain of production and then of conservation of the biological medicinal product. These parameters must be monitored during the final production step of the pharmaceutical form that follows the active substance's production/purification. That step of producing the pharmaceutical form (finished product to be administered to the patient) must also respect the molecular structure, without aggression or degradation.

Then, at this stage of the finished product, the same kind of difficulties arise when a medicinal product known as "biosimilar" and its reference product have different pharmaceutical formulations and/or different presentations.

Any change in the excipient formula may indeed modify the stability profile of the molecule, which will be more rapidly degraded, or will give birth to impurities or degradation products, notably, aggregates.

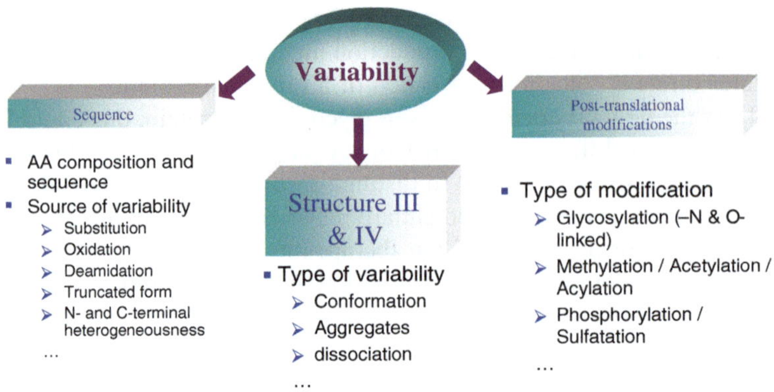

Fig. 4 Molecular structure: a source of variability and heterogeneity

Molecular Complexity

In conclusion of this chapter dedicated to protein molecular characteristics, one should above all remember that biological products are complex structures, not only because of their basic protein structure, but also because of other modifications that they undergo during their maturation, generating a "final form" that is not a "single" and monomolecular entity (as could be expected of a chemical molecule with 99.9 % purity) but rather a complex mix of the same protein molecule under various structurally close isoforms. Here one speaks of intrinsic "microheterogeneity" of a biological substance. The microheterogeneity will be like a digital print of the therapeutic protein which will also be predictive of the protein's activity and tolerance profile when administered to the patient. Products with a close structure that have not been eliminated during the purification steps must be added to this mix of isoforms (when the protein presents a glycosylation profile), as well as impurities brought by the various reagents and different steps of the purification process and the producing of pharmaceutical form (see infra).

This mix will ultimately be considered as *the* product of interest that will be qualified in its global form by clinical data. It has to be considered that, once this "mix" is characterised and validated by clinical use, the producer will have to make every effort to do ensure that, batch after batch, the product delivered to the patient meets the same quality criteria.

Figure 4 attempts to summarize the different sources of variability and the keeping of elementary characteristics needed for the biological activity of a complex structure of a biological substance, such as a protein.

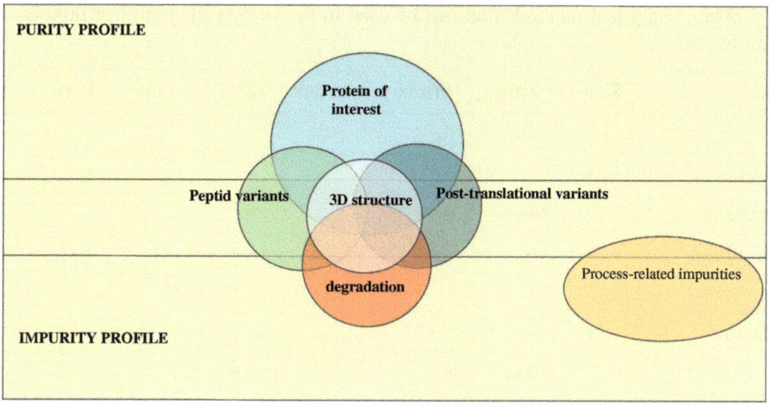

Fig. 5 Schematic representation of the notion of purity/impurity profile

The Analytical Challenge

As explained above, a biological substance has a complex chemical structure and is rarely (if ever) a pure and single molecular entity, but rather a "molecular population" that includes (see Fig. 5):

- the majority molecular form and its variants and isoforms, each carrying an intrinsic biological activity close to the majority molecular form's biological activity (for instance EPO and its isoforms);
- the impurities linked to the product itself, but which practically carry no biological activity;
- the impurities linked to the production process or to the purification scheme.

It is this mix of chemically complex structures that has to be controlled, qualitatively and quantitatively, before considering the release of the medicinal product for its clinical use.

These controls are aimed at verifying, compared to preset specifications, that the medicinal product is the one expected for the patient, at the right dosage, and that the level of impurities conforms to what is expected.

One then understands that the physico-chemical and biological analysis of that complex mix, (to evaluate different distinct molecular properties that are complementary and necessary) cannot be done by a single method, and that, as in the definition of a biological product, "a combination of physico-chemical and biological tests" will be needed to be in a position to extensively characterise the protein of interest in terms of mass, charge, spatial folding and multicatenary or multimeric and finally its post-translational characteristics (quantitative analysis of different isoforms).

There are numerous analytical methods to evaluate the purity and impurity profile of biological medicinal products; among which there are circular dichroism, nuclear magnetic resonance, immunological tests (ELISA, immunoprecipitation, *biosensors*, etc.), biological activity on in vitro models (cell culture) and in vivo (animal models), various chromatography techniques (HPLC *peptide mapping*),

Table 1 Main analytical methods that can be used in the study and control of proteins of other macromolecules

Method	Size	Charge	Primary structure	2°/3°Structure	Purity	Potency
HPLC						
Exclusion size	+++	–	–	++	++	–
Ions exchange	–	++++	+++	–	–	–
Reverse phase	+++	+/–	+++	++	++	–
Electrophoresis						
SDS-Page	+++	–	+++	–	+++	–
IEF	–	+++	+	++	+++	+
W-Blot	+++	–	++	+++	+++	–
Dosages						
Immuno-assays	–	–	+/–	+/–	–	++
Receptors fixation	–	–	++	+++	–	++
In vivo dosage	–	–	+++	++++	+/	+++++

Each method covers all or part of the analysed molecules structural parameters. Data provided by the various analytical methods (adapted from Thorpe R, personal communication)

electrophoresis methods (SDS-PAGE; IEF; CZE), static and dynamic light diffusion mass spectrometry, X-ray techniques, etc.

These methods are in constant evolution in their sensitivity levels, discriminating capacities and specificity. They allow for a continuous improvement of knowledge and understanding of these products. However, these tools reach their limits to exhaustively report a biological product quality profile. This fact has been illustrated several times for biological medicinal products, at different stages of their development (before or after approval), for which variants or impurities have been detected "at a late date." In some cases, these variants or impurities that have always been present in the product, are incidentally discovered because of improved analytical methods and have no clinical consequences. In other instances, impurities have been looked for and detected after pharmacovigilance signals, such as undesirable effects observed in the course of a clinical trial or a long term treatment by these products.

The methods are many and each shows a more or less fine analysis of the various characteristic of the product of interest. The methods may be classified in large families, according to molecular characteristics they are capable to study (Table 1).

Today if many molecular characteristics can easily be investigated and verified (molar mass, glycosylation profile), there are still three-dimensional structural elements that are especially telling criteria, but difficult to analyse in case of complex proteins. The current physico-chemical methods (circular dichroism, near and far UV spectrometry, NMR, etc.) can compare only certain aspects of the protein three-dimensional structure to the given reference protein. Consequently, in the structural analysis of the proteins, some characteristics necessary for the protein biological activity will not be verified by the only physico-chemical methods, and it is routine to perform, on top of physico-chemical tests, so-called biological tests (that involve reactions of the antigen–antibody type, or agonist-receptor, or recognition of an enzymatic site, even tests for pharmacological activity on animals) to verify the molecular global integrity of the molecule.

This analytical challenge also explains that it is sometimes difficult to claim the absence of difference when two molecules are compared. This difficulty to draw a conclusion on a difference or absence of difference explains itself either by inadequate methods or a detection limit reached. In a demonstrated absence of différence, there is a risk to wrongly draw the conclusion that the two compared products are "identical". In this field of comparing molecules, it is important to remember that the absence of proof is not proof of absence; there have been described secondary effects in patients who had received molecules derived from a modified production process, and these molecules had been successfully analysed in search of structural or molecular modifications that the changed process could have induced. It has only been shown after administration to patients that the molecule, produced by a modified process, presented a different behaviour and notably led to the production of antibodies (immunogenicity profile) against the protein of interest, whereas all the analytical parameters concluded for an equal (identical) quality and purity profile.

The Production Process Challenge

An active substance known as "biological" is by nature complex; consequently its production has to make use of processes that are themselves complex, using living materials and reagents with steps of fermentation/cell culture followed by extraction/purification from a complex matrix composed of the chosen expression system metabolism. We shall describe the production process for a recombinant protein, in order to illustrate the "critical" points that will condition the final quality of the protein produced.

It should be briefly recalled that in a process of production called "biotechnological," it is a cell system that ensures the expression of a protein of therapeutic interest. This cell system (bacteria, yeast, insect, plant or mammalian cell) undergoes a genetic modification which consists in the insertion of a foreign gene into the genome of the host cell—one speaks of transgene—coding for the protein of interest. The protein so produced (following the genetic sequence that will have been inserted into the host cell system and after post-translational maturation that the host system is capable of making) will have then to be collected in this culture medium to be purified to the point of getting a protein solution at a degree of purity close to 100 %. It is that purified protein that will be transformed (formulated) into the final pharmaceutical form (most often injectable).

So, to produce a recombinant protein, and get a biotechnological medicinal product, several steps must be identified; each step has an impact on the robustness of the production process, and ultimately upon the quality of the protein of interest:

- developing the cell system expressing the transgene that will have been inserted into its genome;
- implementing the culture of the genetically modified cell;
- "collecting"/harvesting from the production system;
- purification of the protein by different chemical or biological steps in order to get an active substance with a declared and validated purity level;
- pharmaceutical formulation to get the medicinal product in its final form.

We will briefly introduce each of these steps, while putting the accent on the critical point of each of them, since these critical points could fail at any moment and lead to the production of a lesser quality protein or of a protein different from the expected one. In the context of biosimilars, we will see that the process, implemented by a second manufacturer, is therefore one of the essential points to master the quality of a "copy" product.

Developing an Expression System: From Genetic Code to Medicinal Protein

Genes are DNA portions carrying a message that ultimately leads to the production of proteins. They are present in all living creatures' genomes and are sequences of nucleotides (A, T, G and C). Each of these genes' sequence is specific of a protein. The cells' machinery transcribes the genes (DNA) into RNAm which in turn are translated into proteins. These four steps are represented in the following sequence:

DNA → transcription → RNAm → translation → native protein → post-translational modification → mature protein → excretion/secretion → release of the protein into the extracellular medium.

Pre-Production Step, Development of Expression System, Concept of Genetic Engineering

The gene encoding the protein sequence of interest is the first key element of this system since only a correct gene sequence will give rise to the expression of a protein with the right conforming primary sequence.

The gene, most often of human sequence, must be inserted into the cell that will have been chosen as the production host system. As the genetic code is universal, it will be read the same way by all cellular systems of the animal, plant or bacterial kingdom (even as the existence of dominant codons per cell system is known).

This universality is the basis of the production of recombinant[1] therapeutic proteins of human sequence into heterologous host systems (bacteria, yeast, plant, mammalian cell, transgenic animals) to make that host system "produces" a protein of given sequence.

[1] "Recombinant" means a foreign DNA integration into the genome of a host that becomes genetically modified.

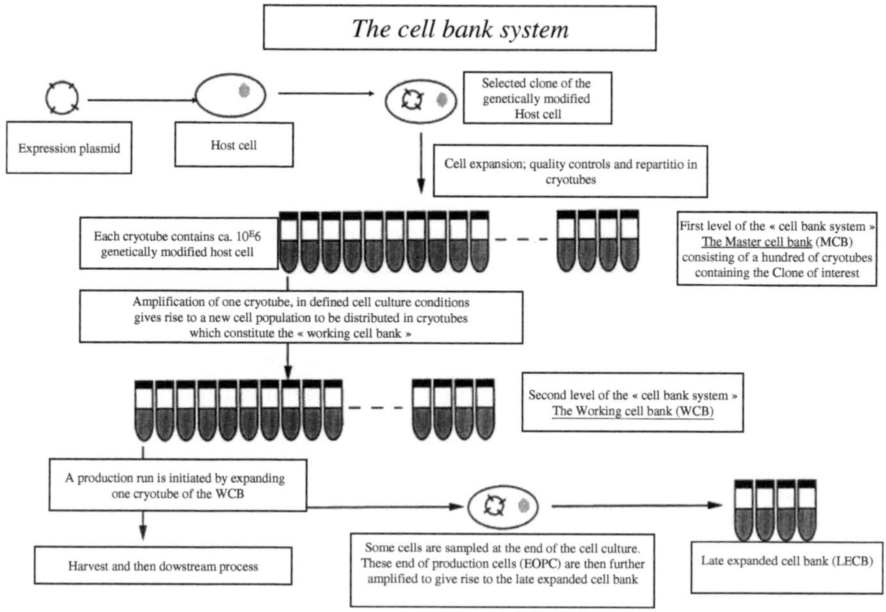

Fig. 6 The two-tier system of cell bank

Biotechnology is therefore aimed at inserting a transgene encoding the protein to be produced in a "host" producer organism named expression system.

This latter is organised in cell banks (master and working cell banks, that system will be described later, Fig. 6) to ensure the production system reproducibility. Cells issued from cell banks are cultured to produce the protein of interest. Then, this protein is extracted from the culture medium, purified, and formulated into a finished product which is the marketed product.

Concept of Expression Cassette and Vector

For the human protein to be correctly expressed by the producer host organism, it is necessary to optimise the gene sequence in using for each amino acid the dominant codon of the species used. The optimised sequence is affixed to a promoter which controls the protein expression by the genetically modified host system. Selecting the promoter depends on the host cell and leads to optimising the expression yield.

At that early stage of "genetic construct," selecting the producer system conditions the nature of the protein produced and related substances which constitute "the active substance of interest."

In other terms, a genetic construct, developed for a given system of production, constitutes a first element of originality and of characterisation of the produced protein.

Any other system of production, developed by a "biosimilar" manufacturer for instance, is by nature different in its design and its construction and therefore in the most intrinsic molecular characteristics of the biological substance produced.

Host Cell and Expression System

Selecting the host cell and the expression system is above all conditioned by the nature of the protein to be produced. In fact, depending on the protein structure complexity and the necessity (or not) of a post-translational modification step, the different cell systems may or may not meet the requirements. Typically there are five types of organisms that may be used in the production of recombinant proteins of therapeutic interest: bacteria, yeast (and fungi), mammalian cells (human cells included), insect and plant cells. To these five main expression systems in vitro, may be added two in vivo expression systems, with transgenic plants and animals, even if the use of these systems is still anecdotic.

It is important to remember over all that, for a given system, the product obtained may notably differ from the others because of its many impurities and the distribution of isoforms and other structural variants that have been stated above. We will briefly discuss some aspects specific to these different expression systems.

Bacteria have been the first in vitro system of production of recombinant proteins, notably *Escherichia coli*. This system combines an easy use and good yields; however, its use is limited to the production of proteins that don't necessitate a post-translational modification, since prokaryotic cells are not equipped in enzymes needed for glycosylation. During the protein production phase, the bacteria will either produce it in "inclusion bodies" inside the cell, or by internal secretion in the periplasmic space. The formation of inclusion bodies produces protein semi-purified fractions that are easily collected by a simple extraction. However, this extraction technique supposes the proteins' resolubilisation in a solvent such as urea, which inevitably implies their partial or total denaturing. This denaturing must then be followed by a renaturing phase of the protein in order for the protein's spatial structure to be regenerated to ensure its biological activity (see supra). This phase of denaturation-renaturation is tricky and if it is not well performed or mastered, may lead to the formation of a more or less consequent fraction of protein molecules with abnormal conformations, which could be responsible for reactions from the recipient organism which don't recognize the denatured protein structures (neo-antigenicity).

Eukaryotic Systems

For complex proteins, necessitating post-translation specific reactions (such as specific conformations, oligomerisation, proteolytic cleavages, phosphorylations or glycosylation reactions) have to be considered production systems able to provide a machinery needed for these maturation reactions, which generally take place in the endoplasmic reticulum and in the Golgi apparatus of eukaryotic cells.

Three levels may be identified within eukaryotic systems:

- lower eukaryotes with yeasts and fungi. These systems are able to make relatively simple post-translational modifications, while preserving their simple culture conditions and a production yield close to bacterial systems'. However, this system remains limited to simple glycoproteins since yeasts produce N-glycosylations rich in mannose residues; strongly immunogenic for humans;
- higher eukaryotes with mammalian cells, but also insect or plant cells. In general these cells are used in industrial production only when microbial or fungal are deemed not adapted to produce the protein of interest. In fact, unlike the systems above, higher eukaryotes are systems more difficult to industrially implement and with a lower yield (as compare to bacteria or yeast), and very constraining culture conditions. Mammalian cells such as ovarian cells of Chinese hamsters (*Chinese Hamster Ovary* (CHO)) are commonly used to produce complex glycoproteins. More recently, human cell lines have been developed and qualified (osteosarcoma 1080 or a line of human embryo kidney HEK293 (*Human Embryonic Kidney*)). These human cell lines have the main advantage of being able to make post-translational modifications of human structures, which could be interesting, notably when the glycosylation profile must be close to the "natural" profile (for instance important for the pharmacokinetic profile);
- at last, a third expression/production system has been recently developed, with plant and animal transgenesis: to have produced by a genetically modified plant or animal, the therapeutic protein, in a tissue (for the plant) or in a fluid (most often milk for transgenic animals) easily accessible for the "harvesting". From this harvest, protein purification and final product formulation follow the production diagram that will be described below.

Concept of Cell Bank

To ensure homogeneity and reproducibility in the protein of interest production, a system of "cell banks" (also designated as two-tiered seed lots systems (Fig. 6)) is set up starting from the initially selected producing clone. For that, the clone of interest (obtained after genetic modification of the host cell line) is put into culture, and the produced population is aliquoted in fractions (a few million cells per cryotube), then cryo-conserved. This first batch of tubes (generally around one hundred of cryo-tubes) constitutes the *Master Cell Bank* (MCB) (since it is directly a product of the selected clone amplified). To ensure the sustainability of this MCB, a *Working Cell Bank* is prepared, by amplifying one or two MCB cryotubes with distribution of cells produced in a working cell bank (WCB) consisting also of hundreds of cryo-tubes, as with the MCB. Each production batch is initiated from a WCB cryotube. It is when, after several production batches have been made, the WCB decreases and a new WCB may be prepared, according to the same process as described above, by taking, in the MCB, one or two cryo-tubes. This "two level" cell bank ensures the sustainability of the cell line that has been built from the same genetically modified cell.

Active Substance Production: Fermentation or Cell Culture Conditions

Once the expression system built up, i.e. the host cell genetically modified and the producer clone set up in a bank, it is time to proceed with the culture of the host cell in order that it produces, in specific and optimised culture conditions, the protein of interest.

These organisms' culture is carried out in fermenters (bacteria and yeasts) or cell culture systems (roller bottle, cytocultivator, hollow fibres, etc.).

It would take too long to detail the different technical solutions that have been developed in order for cells maintained in artificial conditions of production to, not only survive, but above all produce in notable quantities the protein for which they have been genetically modified.

Culture conditions, i.e. the way the host cell is maintained, fed, oxygenated, etc. are determining elements for production yields and the intrinsic quality of the produced protein. To show how important and critical the culture conditions are to obtain a determined quality protein, let's remember that quality and intensity of the glycosylation made by the eukaryotic cells are notably dependent on the culture conditions to which the host cell is exposed. Therefore one may observe that a pH change, a modification of oxygen partial pressure, or a change in the speed of carbon hydrates supply may lead to a modified distribution of glycosylated isoforms.

Thus, for a protein of interest, the production process is a critical step to "harvest" ultimately a bulk product that has to be as reproducible as possible, notably in terms of production impurities, products derived of cell metabolism, and obviously isoforms or molecular variants.

In the context of the production of a protein called "biosimilar", it is understandable why the development of a production process as close as possible to the process used for the reference protein production, and the mastering of this process' reproducibility are key elements to ensure that the biosimilar protein will be, in fact, as similar as possible to the reference protein, in all its structural and physico-chemical aspects. These same questions are evidently implied if the considered production process makes use of transgenic animal or plant, with added elements of complexity and qualification, that we cannot describe here.

Purification

Once the expression and production phase done (whatever the production system, either in vitro or in vivo), it is time for "harvesting" the medium in which lies the protein (bacteria's all body, culture juice of yeast or cells, transgenic animal's milk, etc.) and then to proceed with the steps of extraction and purification of the protein of interest.

The objective of this so-called "downstream process" is to extract from this complex biological matrix the protein of interest and eliminate all the inherent "contaminant" substances, notably those coming from the production cell (DNA and proteins), the used raw materials (reagents, culture media...), and from degradation products. A second purification objective is to eliminate, through dedicated steps, pathogens potentially transmittable and brought in by the different "biological" elements introduced all along the process. Worth mentioning here is the risk of contamination by virus from the cellular system, or other adventitious agents such as the agent responsible for spongiform encephalopathies, etc....

There are numerous strategies to extract and purify proteins, each one presenting a certain level of specificity (in order to select only the protein of interest), of yield (amount of proteins of interest eliminated with effluents), and of maintaining the molecular integrity of the protein being purified.

The purification system may also introduce differences in the protein quality profile, between production batches or between producers in:

- qualitatively and quantitatively selecting isoforms;
- co-purifying different impurities;
- triggering, thanks to more or less drastic purification conditions (notably solvent/detergent treatments typically applied for viral elimination/inactivation), denaturations/degradations of the purified molecules.

At the end of the *downstream process*, there is a said "purified" protein with a level of purity that must be qualified and verified in comparison with preset criteria or specifications (see supra, analytical challenge and quality control strategy).

At this stage of the process, the protein is considered as "the active substance" that can be stored for conservation before producing a pharmaceutical form. In fact, the protein has not yet a "medicinal product" form, meaning a form that could be administered to the patient. It is then time to proceed with the "bulk protein" towards the last step of "making a pharmaceutical form", called the formulation of the finished product.

Towards a Pharmaceutical Form

Making a pharmaceutical form consists of inserting the protein of therapeutic interest in a medicinal form that will be administered to the patient.

As they are proteins, these substances cannot (except for rare exceptions) be administered orally. Therefore the medicinal product to prepare will be administered by injection (subcutaneous, intramuscular or intravenous).

For that, a formulation must be done with excipients able to ensure the protein best possible for dissolving, and maintaining its physical integrity (for instance no aggregate forming) and chemical integrity (no alteration of chemical properties such as oxidation, reduction, loss of groups of glycosylation, etc.), since all these degradation reactions are mainly controlled by chemical conditions (pH, ionic strength, humidity, etc.) and/or physical conditions (storage temperature, liquid or solid state) imposed upon the protein. This step of formulation and excipient selection and of conditions of making a pharmaceutical form allows to ensure the compatibility of the active substance with its final administration environment, but also to guaranty the active ingredient stability inside the medicinal product during its manufacturing process as well as its storage all along the shelf-life.

Thus, this last step must be integrated in the whole process of the medicine's production as a possible source of technical obstacles to achieve a medicinal product of consistent quality.

Among examples in which the medicine's formulation had a consequence in terms of serious undesirable effects, let's mention the case of severe anemia (pure red cell aplasia) due to the formation of antibodies anti-active substance, reported after a change in the formulation of an erythropoietin that was on the market, without noticeable pharmacovigilance signal, for 10 years.

This example, like many others, shows that maintaining the molecular integrity of the protein of interest is resulting from the full mastering of all the process' steps, from the host cell system construction, to the culture of that cell system, then extraction/purification, and finally the making of a pharmaceutical form and compliance with the medicinal product handling and storage conditions.

Conclusion

All along this chapter, we have listed and presented the different scientific and technical elements linked to the very nature of biological substances, their complex molecular structure, their production and purification process, their making into a pharmaceutical form, explaining why there are so many unknowns and so many sources of difficulties in making an exact "copy" of the molecule of reference. It can thus be concluded that the "generics" approach is not applicable to scientifically and sufficiently guaranty the "copy" biological will be of the same quality, safety and efficacy profile as the reference medicinal product against which the "copy" is claimed to be similar.

Only a careful comparison of the two products (the reference and the copy), of their production conditions, of control and storage- reinforced by non clinical and clinical results—will give a chance to the health authorities to guarantee that, within the limits of scientific knowledge, and within the limits of the investigations that had been conducted, there is no possible evidencing of significant differences or potentially generating different efficacy and safety profiles, and that on this basis, the two products are considered as "biosimilars".

Further Reading

- Directive 2004/27/CE du Parlement européen et du conseil (31 mars 2004)
- EMEA/CHMP/31329/05 Annex Guideline on Similar Biological Medicinal Products containing Biotechnology-Derived Proteins as Active Substance: Non-Clinical and Clinical Issues. Guidance on Biosimilar Medicinal Products containing Recombinant Granulocyte-Colony Stimulating Factor
- EMEA/CHMP/31329/05 Annex Guideline on Similar Biological Medicinal Products containing Biotechnology-Derived Proteins as Active Substance: Non-Clinical and Clinical Issues. Guidance on Biosimilar Medicinal Products containing Recombinant Granulocyte-Colony Stimulating Factor (CHMP adopted February 2006)
- EMEA/CHMP/32775/05 Annex Guideline on Similar Biological Medicinal Products containing Biotechnology-Derived Proteins as Active Substance: Non-Clinical and Clinical Issues. Guidance on Similar Medicinal Products containing Recombinant Human Insulin
- EMEA/CHMP/32775/05 Annex Guideline on Similar Biological Medicinal Products containing Biotechnology-Derived Proteins as Active Substance: Non-Clinical and Clinical Issues. Guidance on Similar Medicinal Products containing Recombinant Human Insulin (CHMP adopted February 2006)
- EMEA/CHMP/42832/05 Guideline on Similar Biological Medicinal Products containing Biotechnology-Derived Proteins as Active Substance: Non-Clinical and Clinical Issues
- EMEA/CHMP/42832/05 Guideline on Similar Biological Medicinal Products containing Biotechnology-Derived Proteins as Active Substance: Non-Clinical and Clinical Issues (CHMP adopted February 2006)
- EMEA/CHMP/437/04 Guideline on Similar Biological Medicinal Products
- EMEA/CHMP/437/04 Guideline on Similar Biological Medicinal Products (CHMP adopted September 2005)
- EMEA/CHMP/94526/05 Annex Guideline on Similar Biological Medicinal Products containing Biotechnology-Derived Proteins as Active Substance: Non-Clinical and Clinical Issues. Guidance on Similar Medicinal Products containing Recombinant Erythropoietins
- EMEA/CHMP/94526/05 Annex Guideline on Similar Biological Medicinal Products containing Biotechnology-Derived Proteins as Active Substance: Non-Clinical and Clinical Issues. Guidance on Similar Medicinal Products containing Recombinant Erythropoietins (CHMP adopted March 2006)
- EMEA/CHMP/94528/05 Annex Guideline on Similar Biological Medicinal Products containing Biotechnology-Derived Proteins as Active Substance: Non-Clinical and Clinical Issues. Guidance on Similar Medicinal Products containing Somatropin
- EMEA/CHMP/94528/05 Annex Guideline on Similar Biological Medicinal Products containing Biotechnology-Derived Proteins as Active Substance:

Non-Clinical and Clinical Issues. Guidance on Similar Medicinal Products containing Somatropin (CHMP adopted February 2006)
- EMEA/CHMP/BWP/49348/05 Guideline on Similar Biological Medicinal Products containing Biotechnology-Derived Proteins as Active Substance: Quality Issues
- EMEA/CHMP/BWP/49348/05 Guideline on Similar Biological Medicinal Products containing Biotechnology-Derived Proteins as Active Substance: Quality Issues (CHMP adopted February 2006)
- EMEA/CPMP/3097/02 Note for Guidance on Comparability of Medicinal Products containing Biotechnology-derived Proteins as Drug Substance: Non Clinical and Clinical Issues (CPMP adopted December 2003)
- EMEA/CPMP/BWP/3207/00 Rev.1 Guideline on Comparability of Medicinal Products containing Biotechnology-derived Proteins as Active Substance-Quality Issues (CPMP adopted December 2003)
- Thorpe R et al (2005) Clinical and Diagnostic Laboratory Immunology 12: 28–39

From the Biosimilar Concept to the Marketing Authorisation

M. Pavlovic and J.-L. Prugnaud

Introduction

The concept of similar biological medicinal products similar to a reference biological medicinal product has been recently introduced in the European legislative framework. As it has been stated in this book's introduction, the biosimilar term designates in common language the "copy" concept of a biological medicinal product. The purpose was to open a regulatory route for pharmaceutical companies willing to develop biosimilar medicinal products once the marketing protection of the "reference" biological medicinal product expired.

Several medicines of this particular field are or will have expired patents in the near future, which offers pharmaceutical companies the possibility to develop similar products and to obtain the same therapeutic indications as the reference products.

Even if this strategy could be easily assimilated to the standard generic approach (which, for a chemically derived substance, requires a single demonstration of bioequivalence with the reference product), the generic approach was not considered adequate to establish the quality, safety, and efficacy of biosimilars. That is due to the

Pavlovic article M, Girardin E, Kapetanovic L, Ho H et Trouvin JH, Similar Biological Medicinal Products Containing Recombinant Human Growth Hormone: European Regulation. Horm Res 2008;69:14–21. © 2007. Karger AG, Basel.

M. Pavlovic (✉)
DEMESP Haute Autorité de santé, avenue du Stade de 2,
93218 Paris, France
e-mail: m.pavlovic@has-sante.fr

J.-L. Prugnaud
Agence Française de sécurité sanitaire des produits de santé, Département de lévaluation des médicaments et produits biologiques, Président Commission thérapie cellulaire et thérapie génique, 13 avenue Jean Aicard, 75011, Paris, France

J.-L. Prugnaud and J.-H. Trouvin (eds.), *Biosimilars*,
DOI: 10.1007/978-2-8178-0336-4_2, © Springer-Verlag France 2013

complexity of the biotechnology-derived products themselves as well as their manufacturing process. In most cases, molecular complexity and heterogeneity inherent to biological products do not allow for their full and guarantied characterization.

The quality attributes of the active ingredient are highly dependent on its manufacturing process and any change in the manufacturing process may affect the quality attributes and impact on the safety or efficacy profiles of the product. Therefore the European legislation has provided a specific regulatory framework (called "biosimilar approach") for biological medicinal products similar to reference biological medicinal products. It is applicable to any biological medicinal product, which confers an originality to the European regulation and its unique character. Practically, the biosimilar approach developed in the recommendations for approval application apply to well-characterized recombinant proteins, such as insulin, somatropin, erythropoietin, G-CSF (Granulocyte Colony Stimulating Factor). Other recommendations have been issued for low molecular weight heparins and alpha interferon, or are being drafted for monoclonal antibodies. These guidelines are made by the CHMP (*Committee for Medicinal Products for Human Use*) of the European Medicines Agency (EMA); they are relevant to quality, non-clinical and clinical issues to be developed in order to be addressed for a biosimilar approval application.

Definition of Biosimilars

"When a biological medicinal product similar to a reference medicinal product does not meet the conditions stated in the generics definition, notably because of differences linked to raw material or differences between manufacturing processes of the product and the reference product, appropriate preclinical or clinical studies related to these conditions must be provided." [European guideline 2004/27 art.10 (4)].

In this chapter the general recommendations will be summarized and analysed in relation to the development quality of a biosimilar, followed by those related to preclinical tests necessary before the first human administration. The general recommendations for a clinical evidencing of similarity in terms of safety and efficacy will be particularly developed for biosimilars used in oncology and haematology. These essentially concern erythropoietin and the growth factor G-CS for which the first biosimilars have been put on the market.

Pharmaceutical Authorisation Background

General recommendations appear in a general text on biosimilars, which introduces the concept of biosimilarity and gives a definition of the main principles of biosimilars development in terms of quality, safety and efficacy.[1] A company

[1] Guideline on Similar Biological Medicinal Products. CHMP/437/04 (CHMP adopted September 2005).

developing a medicinal product similar to a reference biological medicinal product must choose the reference among medicines authorized by a complete file with the European Community. The concept of a biosimilar is applicable to any biological medicinal product. However, in practice, demonstrating the similarity will depend on a possible complete characterisation of the product. For that it is necessary to have not only data on physico-chemical and biological properties, but also to know the manufacturing process and its controls. As minor changes in this manufacturing process may alter the product at a molecular level, the biological product safety and efficacy profile depends on the robustness and follow-up of quality issues.

The biosimilar approach takes into account the following points:

- the "standard" generic approach is not considered as acceptable. The biosimilar approach is based on exercises of comparability due to the complexity of biotechnology-derived products;
- exercises of comparability can only apply to highly purified products that may be correctly characterized. It is not always the case, notably for extraction products with biological sources, or those for which only a limited clinical and regulatory experience is available;
- the biosimilar approach is defined by the current recommendations on analytical methods, manufacturing process, and clinical studies conducted for the approval application;
- by definition, a biosimilar product is not a generic product; subtle differences between biosimilar and reference may exist and call for a prior experience before using them. In order to facilitate a later follow-up (pharmacovigilance), patients receiving a biosimilar must be clearly identified.

In the same general recommendations, the same biological reference must be used for the whole program of comparability of quality safety and efficacy studies, in order to ensure that responses during the comparability exercise be obtained with a single comparator, having all along the studies concerning the same form and same dosage, the same types of impurities and variants linked to its manufacturing process. A biosimilar's active substance must be similar in molecular and biological terms to the reference product. For instance an α 2a-interferon cannot be biosimilar to an α 2b-interferon. It is strongly recommended for the biosimilar medicinal product to have the same form, dosage and route of administration as the reference medicinal product. If it is not the case, additional data must be given in the context of comparability exercise to justify these differences. Any difference between biosimilar and reference must be justified by appropriate study, case by case. A consultation with regulatory authorities is recommended for discussing these approaches.

Quality Control Approach

Biosimilars are biological products developed according to their own manufacturing process. Scientific data coming from pharmacopeia' monographs or published in the literature on the reference biological medicinal product are considered as limited in order to establish the similarity between biosimilar and reference at the active substance and finished product levels, for they are not relevant enough. Only a comparability exercise will allow the evaluation of similarity in terms of quality, safety and efficacy. Based on a complete quality dossier combined with sensitive analytic tests, the comparability exercise at the quality level allows the reduction of the number of non-clinical and clinical studies, compared to a complete approval application file.

A complete quality file, comparable to the file required for the reference medicinal product approval, is always required for biosimilars approval applications. It is completed with quality, non-clinical and clinical comparability data between the reference medicine and the biosimilar medicine.

Biosimilars' Manufacturing Process

The biosimilar is defined by its manufacturing process specific to the active substance and to the finished product (as for the reference medicinal product). These processes must be developed and optimized according to current regulatory recommendations, covering aspects of the molecular expression system and of the production cells, culture, purification, viral protection, formulation excipients, interactions with primary packaging materials, as well as their possible consequences upon the finished product characteristics. Besides, every medicine is defined by its molecular composition, which is itself defined by its manufacturing process which introduces its own impurities. For these reasons, the biosimilar is defined by:

- the molecule itself, including variant products and impurities;
- the manufacturing process which may play upon molecular characteristics and impurities.

The company that develops the biosimilar must master all these issues in terms of reproducibility and robustness of the processes involved. It is recommended that clinical data in the comparability exercise be obtained with the biosimilar manufactured according to the final manufacturing process that will be used for batches to be marketed. Otherwise "bridge" studies will be needed.

Quality Comparability Exercise

Quality issues are essential for a biosimilar and their potential impact on safety and efficacy must always be evaluated. A step by step approach is recommended in order to analyse and justify any difference in the quality attributes between biosimilar and reference. It is not demanded that the quality attributes be identical as minor structural differences may exist for the active substance, due to the post-translational modifications' variability or differences in impurities profile.

These may be acceptable but must be justified, notably in terms of their possible impact upon safety and efficacy of the finished product.

Analytical Methods

Characterisation studies must be conducted according to regulatory current recommendations concerning the active substance and at the same time the final product to demonstrate that the biosimilar quality is comparable to that of the reference. The analytical methods must be chosen according to the product's complexity and must be able to detect differences between biosimilar and reference. The comparison is done with validated analytical methods assessing composition, physical properties, primary and higher degree's molecular structure, different forms related to post-translational modifications, and biological activities. Several biological tests are needed; they use various approaches in order to compare the biosimilar's and reference's biological activity. Activity expression must be stated in international units, if an international standard exists.

A biosimilar's derived products and impurities must be identified and compared to its reference's using current available techniques. Stress studies are used to show specific degradations (*i.e.* oxidation, dimerization) and accelerated stability studies lead to profiles of stability that can be compared between biosimilar and reference.

Impurities related to the manufacturing process (proteins and DNA [deoxyribonucleic acid] of the host cell, reagents, purification impurities) are specific and depend on the manufacturing process of each product. Because of this the comparability exercise may not be applied in an absolute manner. However the biosimilar, as the reference product, must meet the same level of requirements described in the recommendations on biotechnology-derived products quality.

Specifications

As with any biotechnology-derived product, the specifications are based on a selection of tests depending on the given product. The rationale for fixing the limits of acceptation criteria must be described and developed following the same approach as for any biological medicinal product. Each acceptation criterion must be established and its justification must be based on batches used in non-clinical and clinical studies, on batches produced in a reproducible way, and on data coming from comparability exercise (quality, safety, efficacy).

To fix specifications, the company that files the marketing authorisation application must use a global reasoning: this application is based on experience acquired from the product being developed and its reference medicine. Data must show, if possible, that the limits of a given test are not wider than the variability deviations observed with the reference medicinal product.

Conclusion on Quality

The quality aspect in a biosimilar's development is essential. It is on that aspect that mostly lays the demonstration of similarity between biosimilar and reference. The quality file of a biosimilar must contain the two following demonstrations:

- characterisation and production full studies, on active substance and on finished product;
- a comparability exercise to evaluate the quality and similarity of the biosimilar and the reference. These studies have to be interpreted in the context of safety and efficacy comparable between biosimilar and reference. In the biosimilar approach, if data concerning quality are crucial, they have, however, to be completed with data coming from non-clinical and clinical comparative studies, more limited than those required for the development of a brand new medicine.

Non Clinical and Clinical Aspects

Quality, safety and efficacy are key issues that must be followed during a medicine's whole life. For a typical chemical medicine, pharmaceutical development is well-defined. It includes data to document the pharmaceutical quality and is completed by preclinical, called toxicological data, before the first human administration. Clinical development requires data concerning a proof of concept, dose evaluation and demonstration of efficacy in pivotal studies conducted in the medicine's target population. Based on quality, preclinical and clinical studies, stored during its development, the medicine may be ready to be filed in order to get a Marketing Authorisation (MA). As for chemical medicines, the application biosimilar approach necessitates the development of the manufacturing process (for the active substance and finished product), and the demonstration of safety and efficacy through non-clinical and clinical studies. However, as the reference biological medicinal product has been already approved and used for many years in the European Union, its data are available in the public domain. Consequently, a biosimilar development calls for less non-clinical and clinical data than a new medicine; some of this data may be taken as given with the reference product and be used as "support" data in the biosimilar file. Thus, if the reference is approved in several clinical indications, and its mechanism of action is the same in all approved indications, then it is possible to assume that there is a "therapeutic similarity" between reference and biosimilar and to extrapolate the biosimilar efficacy demonstrated for one indication to other indications of the reference medicinal product.

Preclinical Approach

Preclinical studies are comparative and generally include in vitro studies of receptor bindings and tests on cells already found in quality data provided for the biological activity evaluation. These studies can establish the comparability in terms of their mechanism of action between compared products and identify causality factors in case comparability could not be established. In vivo studies on relevant animal studies must be added, while taking into account the regulatory guidelines in force.

Preclinical study has to evaluate, when the animal model allows it:

- activity in connection with the pharmacodynamic effect relevant to the clinical application;
- non-clinical toxicity determined with a single and repeated dose; it is not necessary to have toxic dose finding studies, as they are already known. Measurements in toxicokinetics include the determination of the level of antibodies with the study of crossed reactions and of the neutralisation capacity; the studies must last long enough to show any difference relevant in terms of toxicity and/or immune response between the biosimilar and the reference product;
- if necessary, local tolerance comparative studies.

Other routine toxicological tests (safety pharmacology, reproductive tests, mutagenicity, carcinogenicity) are not necessary. The preclinical studies program is a limited program due to the fact that the toxicology data are known for the reference medicinal product and it is not necessary to repeat all the studies to know the biosimilar.

Clinical Approach

The exercise of clinical compatibility is done step by step; it generally starts with pharmacokinetics and pharmacodynamics studies in healthy volunteers. These studies are followed by efficacy and safety comparative studies. In most cases, the clinical efficacy studies are conducted to demonstrate a therapeutic equivalence between the biosimilar and the reference in a population of patients chosen for the most sensitive to the studied medicinal product effects in order to evidence any difference that could be exist between biosimilar and reference. However, even if efficacy is demonstrated through a therapeutic equivalence test, a biosimilar tolerance may differ from the reference's if there are differences in terms of quality attributes not apparent or difficult to analytically demonstrate. These differences may have unpredictable clinical consequences, and a biosimilar clinical tolerance must be continuously evaluated before and after its marketing authorisation.

During the evaluation of clinical tolerance, a special attention has to be paid to immunogenicity, because patients may develop against the biosimilar as against any recombinant protein in some circumstances; these antibodies could have clinical consequences. The immunogenic potential of a biological medicinal product differs between products and depends on several factors like the active substance's nature and structure, impurities, excipients of the medicine, manufacturing process, route of administration, and target population. These differences may compromise the product in vivo behaviour, with, as a consequence, undesirable effects for the host that may minimize the intended clinical effect with potentially lethal reactions.

Different approaches based, for instance, upon the response of the epitope to Human Leucocyte Antigen (HLA) polymorphism, or the immunological response studied in relevant animal models, may be used to evaluate a biosimilar immunological profile. However, if these responses are important to identify the antigenic profile, they are not predictive of the immunological response to the biosimilar in vivo. Evaluation of a biosimilar antigenic profile in patients is complex because of the difficult measurement of antibodies' level (unavailability of immune serums, absence of appropriate standards, interference of endogenous proteins, limits of analytical methods, etc.) Similarly, the simple comparison of products of the same therapeutic class, although interesting on a theoretical level, is not enough and may be the source of misinterpretation.

Overall, the decision to put a biosimilar on the market is made if its efficacy is similar and its immunogenic profile is at least comparable or improved in comparison to the reference product. However, this decision is made on limited data. The comparability program may disclose substantial differences in terms of immunogenic profiles but is probably unable to detect minor differences and rare events. For that, clinical trials complemented by a pharmacovigilance program are essential for evaluating a recombinant protein's safety in patients. Some undesirable effects are very rare and require a follow-up during the medicinal product's whole life; this is particularly true for biosimilars.

Recommendations in Onco-Haematology

Hematopoietic Growth Factor (rG-CSF)

The file of a biosimilar of Recombinant Granulocyte Colony-Stimulating Factor (rG-CSF) that positions itself as similar to a medicine already approved in the European Community and whose patent has expired must demonstrate its comparability in terms of non-clinical and clinical quality with the reference product.

The human G-CSF is a protein made of 174 amino acids with an O-glycosylation site on a threonine. The recombinant protein obtained in *E.coli* is not glycosylated and presents an additional terminal methionine. The rG-CSF protein has a free cysteine and two disulfide bonds. The medicines rG-CSF obtained by expression in *E. col* (*Filgrastim*®) and in CHO [Chinese Hamster Ovary] (*Lenograstim*®) are clinically used for several indications:

- reduction of the duration of a severe neutropenia after a cancer chemotherapy or myelosuppressive treatment followed by a bone marrow transplant;
- mobilisation of hematopoietic stem cells in peripheral blood (Peripheral Blood Progenitor Cell [PBPC]);
- treatment of severe congenital, cyclic or idiopathic neutropenia
- treatment of persistent neutropenia in Human Immunodeficiency Virus (HIV) patients.

Doses Vary with Indications

G-CSF acts on target-cells through a membrane receptor. Only one soluble isoform that attaches itself to the extracellular part of the receptor is known. The extracellular binding domains of known isoforms are identical. Consequently, G-CSF effects are mediated by only one class of receptors.

The approval application and marketing of a G-CSF biosimilar require comparative studies of non-clinical and clinical quality.

Non Clinical Program for rG-CSF

The non-clinical program includes:

- comparative pharmacodynamic studies:
  - in vitro at receptor level on adapted cellular models, to measure biological activity;
  - in vivo on neutropenic and non neutropenic rodent models, in order to compare the biosimilar effects to those of the reference;
- toxicology studies with a single or repeat dose to a relevant species for at least 28 days.

  Other routine toxicity tests are not required.

Clinical Program for rG-CSF

The clinical program to compare biosimilar to the reference product includes:

- pharmacokinetics studies in crossed single dose for the different routes of administration (subcutaneous, and intravenous) in healthy volunteers. Studied parameters include the area under the curve (AUC), the C max and T ½ with an evaluation performed according to bioequivalence general principles;
- pharmacodynamics studies—the absolute number of neutrophils is the pharmacodynamics marker the most relevant for G-CSF activity. The pharmacodynamics study may be done during the pharmacokinetics with a dose selection in the ascending linear part of the dose–response curve; repeat dose studies may be necessary. CD34+ level is a secondary pharmacodynamic parameter;
- the clinical model suggested for efficacy clinical studies is the prophylaxis of sever neutropenia after cytotoxic chemotherapy in a group of patients homogenous in terms of tumour type and in terms of programmed and validated chemotherapies according to the tumour stage. A two-arm study comparing

biosimilar and reference is recommended with the measurement of frequency and duration of neutropenia as the efficacy main criterion. The company must justify the clinically acceptable difference in the sever neutropenia duration (ANC <0,5 × 10/L) between the biosimilar and the reference. This evaluation will be done during the first cycle of chemotherapy;

- G-CFS effects are mediated by only one class receptors and the results of clinical comparability obtained on the model may be extended to other indications of the reference product;
- clinical safety must be evaluated from a cohort of patients who have received repeat doses of biosimilar, preferably during the comparative phase of the clinical trial. The total exposure of patients must correspond to the normal exposure of the conventional treatment with a corresponding number of chemotherapy cycles. The duration of the study must not be shorter than six months and must integrate immunogenicity data. The number of patients must be sufficient for evaluating the secondary effects including bone pains and biological parameters;
- a strengthened program of pharmacovigilance must be implemented with a risk management plan. The two must take into account that immogenic events are rare but serious in patients with a chronic administration.

Erythropoietin

Human erythropoietin (EPO) is a 165 amino acid-glycoprotein produced in the kidney, that stimulates the production of red blood cells. The medicine is obtained from recombining DNA technology in mammal cells able to express a glycosylated protein.

The recombinant protein has the same sequence as the natural protein but differs by the number and types of isoforms. The protein's glycosylation influences efficacy and safety including the protein's immogenicity.

Erythropoietin based medicines are indicated in various conditions such as anemia in patients suffering from chronic renal insufficiency in patients treated by a cancer chemotherapy inducing an anemia, and also in some programs of autologous transfusions differed in order to increase the number of autologous blood donations. The active substance's mechanism of action is the same for all indications currently approved but the doses to get the desired response vary a lot and are generally higher for cancer indications. The medicine is injected by IV or SCD.

As it is generally well-tolerated, EPO allows a range of therapeutic concentration relatively wide. The hemoglobin content reached allows a control of the bone marrow stimulation and consequently of doses and periodicity of the treatment. The hemoglobin content increase varies considerably between patients and depends on numerous factors like dose and administration rhythm but also the level of iron in the body, basal content of hemoglobin and endogenous erythropoietin, and concomitant treatments or patient's underlying condition, such as inflammation.

The pharmacodynamic response must be under control to avoid serious undesirable effects like high blood pressure and thrombotic complications. Cases of Pure Red Cell Aplasia resulting from the production of anti-erythropoietin neutralising antibodies have been observed, mainly in patients with chronic renal insufficiency and treated with sc injections. Stemming from the fact that usually these antibodies' production is a very rare event, clinical studies for pre-marketing authorisation do not identify these events. Other considerations have to be taken into account for erythropoietin approval applications that are their possible angiogenic action and tumour promoter. Thus the study population selection is particularly important.

The approval application files for a new biosimilar erythropoietin involve the demonstration of comparability with the reference product in terms of quality, safety and efficacy.

Non Clinical Program for EPO

The non-clinical studies include:

- pharmacodynamic comparative studies:
  - in vitro to evaluate the absence of altered response on receptors, with tests of binding to receptors or with cellular proliferation tests. Some tests come from quality comparative studies;
  - in vivo to evaluate the erythrogenic action on relevant animal models. Information on the erythrogenic activity may be obtained through toxicity studies with repeat doses or specifically with a methodology like the one described in on mice in the European Pharmacopeia (Normocythaemic Mouse Assay);
- single and repeated dose toxicity studies on a species relevant to rats. The studies must last at least four weeks and include a toxicokinetics evaluation;
- local tolerance studies, notably with repeated doses with subcutaneous injections.

  Other routine toxicity tests are not required.

Clinical Program for EPO

The clinical program is comparative between copy and reference; it is made of pharmacokinetics studies in crossed single dose for the different routes of administration (subcutaneous, and intravenous) in healthy volunteers. The dose has to be chosen in the sensitive part of the dose–response curve. Studied parameters include the area under the curve (AUC), the C max and T ½. The bioequivalence margins must be beforehand defined and justified;

- the pharmacodynamic parameters must be preferably studied during pharmacokinetics. In single dose studies, the most relevant parameter is the number of reticulocytes, for it is a pharmacodynamic marker of erythropoietin's activity. However this marker does not substitute for efficacy, since it is not directly correlated with hemoglobin level;
- clinical biosimilarity must be demonstrated by comparative clinical studies powerful enough, randomised and in parallel groups between biosimilar and

reference. As pharmacokinetics and efficient doses differ between IV and SC routes, studies must be conducted on each mode of injection. The studies may be conducted either separately for each route, or for one route with appropriate "bridge" data for the other route. Double blind studies are preferable in order to avoid any bias;

- sensitivity to erythropoietin is better for patients who have a deficit in endogenous erythropoietin than for patients without a deficit. Patients with chronic renal insufficiency without major complications will be preferred as a model population for the biosimilar's clinical trials. The other possible anemia causes will be excluded from the comparability studies. The populations in dialysis and predialysis shall not be mixed, as the doses needed to maintain the hemoglobin level are not the same;
- it is possible to demonstrate efficacy's similarity through different options and recommendations. Two different clinical trials are conducted; the trials may combine a phase of anemia correction by sc injections (for instance for a predialysis population) and a maintenance phase by iv injections (for instance for an haemodialysis population). During the correction phase, the dynamic response and the dose may be determined by carefully checking on the safety profile of biosimilar's patients. This phase may include treatment-na patients or patients already on treatment after a three-month treatment free period. In the maintenance phase, patients must have an optimal titration on reference product for at least 3 months. After this period, they are randomised between biosimilar and reference product, while keeping the erythropoietin prerandomisation dose, as well as the periodicity and the administration route. For the correction phase, the responder rate or the change in hemoglobin level may be chosen as a primary endpoint of clinical activity. Anyway, dosing erythropoietin remains the trial's secondary endpoint. A four-week evaluation period is necessary for a study lasting 5–6 months, for the correction phase as well as for the maintenance phase. The studied must be designed according to a methodology fit for evaluating the equivalence between the two products; another approach is to conduct a comparative efficacy study for one route of administration and to provide, for the other route, data resulting from "bridge" studies comparative of PK/PD in single dose and multiple dose, conducted in a population sensitive to erythropoietin (for instance healthy volunteers). The PK/PD study in multiple doses must last four weeks minimum, with a fixed dose of EPO with a primary endpoint fixed on the evolution of hemoglobin level; in all cases of immunogenicity comparative data are required for sc route. In comparative sc route studies, a total duration of twelve- months' treatment is required.
- the clinical safety data are generally sufficient to provide a satisfactory data base for pre-marketing authorisation. The undesirable effects' follow-up notably includes high blood pressure and its possible aggravation and thromboembolic events. The company must file immunogenicity data coming from a 12 months' period for the biosimilar's application file. A validated test sensitive to detecting early and late antibodies must be implemented during correction and maintenance phases. Searching for the presence of neutralising antibodies or Pure Red

Cell Aplasia episodes during the pre-authorisation phases is crucial; it must be complemented by an adequate post-MA follow-up. The data allowing to demonstrate a clinical similarity come from the comparative trial on the population considered the most relevant (chronic renal insufficiency), both in iv and in sc (on a number of patients big enough, as it is commonly accepted that the sc route is more immunogenic that the iv route);

- the typical pharmacovigilance program is completed by a risk management plan notably taking into account rare and serious secondary effects like Pure Red Cell Aplasia of immune origin and the EPO's potential action of tumour promoter;
- as EPOs' mechanism of action is identical for all approved indications for the reference product, and since there is a known EPO receptor, the demonstration of efficacy and safety in the chronic renal insufficiency population makes possible the extrapolation to other reference medicine's indications for the same route of administration.

Conclusion

The biosimilar approach based on an exercise of comparability with preclinical and clinical data, in addition to quality data, allow pharmaceutical companies to file a shortened file (compared to a standard complete file required for a new biotechnology-derived medicinal product) in order to obtain the MA of a biological product similar to the reference biological product; it is called a "biosimilar." It is more necessary to establish a given level of similarity in terms of quality issues than in terms of safety and efficacy, for the biosimilar to be approved in one or all indications of the reference medicine's. Biosimilars are above all biological medicinal products characterized by their own quality profile. The long-term consequences of possible differences between biosimilar and reference are not well known because the clinical trials, conducted over a short period, are designed to demonstrate the equivalence of efficacy and pharmacodynamics. The long-term safety profile will be known only after several years of these products' use. Because of that fact, a biological medicinal product cannot be substituted by a biosimilar medicinal product (as for standard generics) before having collected long-term data on efficacy and safety of the product in all populations to be treated. Currently, in France, the substitution of a biological medicinal product by a pharmacist is not possible. Only a medical prescription in controlled conditions may allow the substitution of a reference biological medicine by a biosimilar.

Further Reading

- Directive 2004/27/EC du Parlement européen et du conseil modifiant la directive 2001/83/EC instituant un code communautaire relatif aux médicaments à usage humain (31 mars 2004)

- EMEA/CHMP/437/04 Guideline on Similar Biological Medicinal Products (October 2005)
- EMEA/CHMP/BWP/49348/05 Guideline on Similar Biological Medicinal Products containing Biotechnology-Derived Proteins as Active Substance: Quality Issues (CHMP adopted 22 February 2006)
- EMEA/CPMP/ICH/5721/03 ICH Topic Q5E Comparability of Biotechnological/Biological Products (CHMP adopted December 2004)
- EMEA/CPMP/ICH/302/95 ICH Topic S6 Step 4 Note for Preclinical Safety Evaluation of Biotechnology-Derived Products (CHMP adopted September 97)
- EMEA/CHMP/42832/05 Guideline on Similar Biological Medicinal Products containing Biotechnology-Derived Proteins as Active Substance: Non-Clinical And Clinical Issues
- EMEA/CHMP/BMWP/31329/05 Annex Guideline on Similar Biological Medicinal Products containing Biotechnology-Derived Proteins as Active Substance: Non-Clinical and Clinical Issues. Guidance on Similar Medicinal Products containing Recombinant Granulocyte-Colony Stimulating Factor (CHMP adopted 22 February 2006)
- EMEA/CHMP/BMWP/94526/05 Annex Guideline on Similar Biological Medicinal Products containing Biotechnology-Derived Proteins as Active Substance: Non-Clinical and Clinical Issues. Guidance on Similar Medicinal Products containing Recombinant Erythropoietins (CHMP adopted 22 mars 2006)

Immunogenicity

J.-L. Prugnaud

Introduction

Immunogenicity may be defined as the power of an antibody to induce an immune response in a given individual in appropriate conditions. An antibody may be antigenic without being immunogenic if, at least in certain conditions and in certain subjects, it is able to induce a response in binding itself in a specific manner to immunoreceptors. Nevertheless, the antibody injection will not trigger in the given individual and given conditions an immune response. Thus, in the definition of immunogenicity there is a quantitative notion linked to circumstances that is found again in the fact that some antibodies are very much immunogenic and others are little [1].

All proteins are potentially immunogenic. The therapeutic proteins can therefore always trigger an immune response when they are injected into the body. This immune response may have more or less serious consequences, from a simple tolerance reaction to antibodies, up to therapeutic inefficiency when the antibodies are neutralising. Antibodies produced against therapeutic proteins like erythropoietin (EPO) [2], hematopoietic growth factors (GM-CSF) [3] and thrombopoietic/megakaryocyte (TPO/MGDF) [4] may have big consequences to the point of blocking not only the exogenous protein's activity but also of the endogenous protein with the serious complications inherent in these actions.

The production of antibodies against biotechnology-derived proteins like insulin, factor VIII or IX, or interferons, does not have the same serious consequences and doctors go on with the treatments in presence of the antibodies, adapting the doses of therapeutic protein [5–7].

J.-L. Prugnaud (✉)
Prugnaud Consulting, 13 ave Jean Aicard, 75011 Paris, France
e-mail: jlprugnaud@gmail.com

J.-L. Prugnaud and J.-H. Trouvin (eds.), *Biosimilars*,
DOI: 10.1007/978-2-8178-0336-4_3,

The consequences of antibodies produced against monoclonal antibodies have been observed since their first use, particularly when these proteins are derived from animal or bacterial proteins. The reactions observed could be of a general order such as systemic reactions, during these products' injection, local reactions or reactions of acute hypersensitivity (generally not due to the antibodies). The immune reactions of anaphylactic type or allergic reactions have been observed relatively often and are now rarer because of the better purification of proteins produced by recombinant DNA technology and the humanisation of protein skeletons of monoclonal antibodies [8–10]. The production of neutralising antibodies may correspond to several types of mechanisms like the direct bonding to a biological activity site or to a site which is not in direct relation but impedes its activity by a changed structural conformation. The non-neutralising antibodies bind to the therapeutic protein site without affecting the biological activity site. If they don't directly neutralise the biological target, they may change the medicine's bioavailability by increase of the clearance of the bonding complex made with a result identical to that of biological activity neutralisation [8].

Whatever the nature of the antibodies produced, the immune responses generated by therapeutic proteins pose a significant problem of safety and efficacy for the authorities in charge of evaluating and approving the marketing of biological medicinal products. Recommendations [11] concerning the evaluation of the biosimilar's immunogenic profile comparatively to the references have been published. These recommendations are based on a multifactor approach taking into account the mechanisms involved; the different factors that may be part of the immune response, and the level of expression of antibodies and the possible rarity of the response observed. It is compulsory for the companies to evaluate the biosimilar's immunogenic risk, case by case, to ensure its safe use. No specific method is recommended, taking into account the variability and multiplicity of factors involved, but actions to take must be identified before clinical trials start, as well as evaluations that will be performed during pivotal clinical trials and evaluations that will be done after marketing of the product, notably within the framework of a risk management plan.

To measure the consequences of the risk linked to immunogenicity during the therapeutic use of biosimilars, immune mechanisms at play must be examined as well as factors having an influence on immunogenicity.

Immune Mechanisms

It is currently accepted that antibodies' formation against therapeutic proteins may be done by two canals: a typical immune response comparable to the one directed against foreign proteins, and a response of tolerance breaking down vis-à-vis the "self-proteins" [9].

Typical Immune Response

In general, a typical immune response occurs after the administration into the body of a foreign protein. Antibodies are produced according to the following classic sequence:

- antigen-presenting cells capture the foreign protein and cleave peptide bonds, degrading proteins into peptides that associate then to class II molecules of the histocompatibility major system or HLA system in humans. The lymphocytes T CD4+ recognise the peptides presented by the class II HLA molecules and activate the lymphocytes B that produce the antibodies specific to these peptides derived from the foreign protein. There is an "immunologic memory" phenomenon that translates into "effector memory" and by a "central memory". The effector memory leads to a rapid destruction of the foreign protein when it is reintroduced in the body, when the central memory is the capacity for producing antibodies and effector T cells more rapidly and for a lower dose of antigen than during the primary response;
- the classic immune response is observed when foreign proteins of animal, microbial or plant origin, are administered to humans. The antibodies formation is usually quick, from a few days to one or several weeks, often after a single injection. The antibodies are most often of the neutralising type. They may persist for a long period.

A classic type immune response may be observed with human proteins produced with the recombinant DNA technique in patients presenting an innate immune deficit. In fact, the production of neutralising antibodies has been observed with the recombinant coagulation factor VIII and with the recombinant growth hormone in such patients.

Immunologic Tolerance Breakdown Response

Numerous recombinant human proteins are homologous to endogenous protein structures also called "self-proteins". After injection, these recombinant proteins don't induct an immune response, as the body is "tolerant" to these molecules considered as self-proteins. However, antibodies directed against these recombinant proteins can be observed in an immunologic tolerance breakdown. The formation of antibodies by this process is slow; it often appears in patients who had received chronic treatment for months. Generally, these antibodies disappeared when the treatment ended. The exact mechanism of immunologic tolerance breakdown is not known. Physiologically, there are lymphocytes B which recognize self-proteins. There lymphocytes B are called "self-reactive". When a self-reactive lymphocyte B interacts with an epitope present in a repetitive form, a cross-linking of receptors B occurs, which leads to an activation of self-reactive lymphocyte B, and the synthesis of antibodies. This mechanism could be related to the immune system of lymphocytes B recognising microbial type patterns, independently from the self/non-self discrimination exercised by these cells. An illustration of this mechanism of tolerance immunologic breakdown would be the immune system response directed against the repetitive protein structures such as aggregates. Some aggregates demonstrate a structure analogous to repetitive structures. This structure would induce an initial activation of lymphocytes B with production of IgM type antibodies. The occurrence of this IgM/IgG transformation, elucidating the IgG synthesis specific of aggregates, is done according to a mechanism unknown to this day. Some studies suggest that aggregates are internalised after reacting with the receptors of lymphocytes B. After internalisation, lymphocytes B would synthesise cytokines capable of activating other lymphocytes B. Inversely, the authors suggest the existence of a different mechanism involving auxiliary T lymphocytes, called T helpers. However, no published data has described the presence of specific T cells in patients exhibiting antibodies directed against therapeutic proteins. Finally, the absence of association with HLA haplotypes and the absence of immunological memory suggest a mechanism independent from T lymphocytes.

Factors Influencing Immunogenicity

Therapeutic proteins immunogenicity is influenced by different factors. Some concern the protein's very structure, how to produce it; with its purification degree, its formulation in order to make a medicine out of the therapeutic protein, the treatment type and the patients' characteristics, plus other factors possibly not known.

Structural Factors

Proteins are complex molecules with a primary, secondary and tertiary structure.

Primary Structure

Changes within the primary structure may be the cause of an immunogenic reaction. Several cases are well known and published in the literature:

- changing an insulin amino acid is enough to lead to a strong immunogenic response, whereas two amino acids inversion only leads to a pharmacokinetic change;
- the homology degree of a recombinant protein with the natural protein may explain an immunogenic reaction but the well-known case of recombinant human α interferon that shows 10–23 amino acids different from human α interferon (homologous for about 89 %) does not lead to immunogenicity exacerbation. Foreign proteins like streptokinase, salmon calcitonine etc., are known for inducing classic immunogenic reactions in patients;
- reactions of oxidation or deamidation of amino acids are known for triggering an immunogenic reaction by forming new epitopes. It is the example of human recombinant α interferon with one methionine, oxidised because of a modification of the purification process, that has led to non neutralising antibodies formation and which, returning to the initial process, has stopped being immunogenic;

- the modification of stability characteristics of a protein with aggregates formation may have significant consequences in terms of immunogenicity by tolerance breakdown of the immune system.

Glycosylation

Glycosylation is also an important factor in therapeutic protein immogenicity. Glycosylation is a post-translational modification, cell and species dependant. It is well-shown that proteins with a human structure produced in eukaryote non-human cells give immunogenic human responses. Similarly, the glycosylation level plays an important role. A β interferon produced in *E.coli* is more immunogenic than that produced by mammal cells since the latter's glycosylation, helping its solubility, decreases the formation of immunogenic aggregates. A protein's glycosylation can decrease immunogenicity by masking antigenic sites.

PEGylation

Some recombinant proteins are "pegylated" in order to modify their pharmacokinetics. PEGylation of a protein is the process of attachment of polyethylene glycol (PEG) chains to the skeleton of the protein. The result is a decreased total half-life of the protein, proteolytic enzymes protected, and sometimes masked immunogenic sites. The new proteins so obtained differ by their conjugated structure, their molecular size and their spatial conformation (linear, branched, or multi-branched chains). Often does PEGylation lower the protein's immunogenicity, probably through multiple mechanisms related to blocked antigenic sites, solubility improved, and lower administration frequency of the therapeutic protein. In general a branched PEG protein is more efficient than a linear PEG protein because of improved immunological properties. However, examples have been published in the literature of PEG proteins more immunogenic than non-PEG (PEG-rhMGDF and rh-TNF coupled with PEG dimer).

Secondary and Tertiary Structures

The significance of protein spatial conformation is well-known for its biological activity as well as tits stability. Partial modification of spatial conformation may occur after shear, by shaking, or by temperature modification (for example: temperature rise or freeze/thaw cycles). Aggregation phenomena may expose new epitopes to the protein's surface for which the immune system is intolerant. That leads to a standard immune response.

In other conditions, protein aggregation may lead to presenting a multimeric antibody, which is known for not triggering B lymphocyte tolerance breakdown. This is why, in a therapeutic protein's analysis, it is important to look for the presence of aggregates and to limit their presence to a low level in the formulated medicine.

Impurities and Other Production Contaminants

Therapeutic proteins obtained through recombinant DNA technology are produced in various cellular systems where production-linked protein impurities originate. These proteins are called host cell proteins (HCP), considered "non-self" proteins by the immune system and may lead to antibody formation by the standard immune mechanism. If by nature anti-HCP antibodies don't neutralize the biological activity of the therapeutic protein of interest, they can nevertheless have consequences in terms of general effects including skin reactions, allergies, anaphylaxis or a serum sickness. Other contaminants, such as impurities, coming from chromatographic column resins or from enzymes used for refining therapeutic proteins' purification, may be found as traces in the finished product. Some impurities may be released by some compounds used for the capping process. These impurities may play the role of amplifier for the immune response, even if they are not able to initiate an immune response themselves.

Producing therapeutic proteins is a complex process that, with time, necessitates change, sometimes major, to keep the production system at an even level of efficacy. It is important to verify that the same levels of quality and safety are maintained. Several examples exist on the modification of a recombinant protein's immunogenicity with time, involving production bacterial cells endotoxin level. DNA G-C patterns from bacteria or degraded proteins are able to activate Toll-like (Toll-like receptors are a class of proteins playing a key-role in the innate immune response. They are transmembrane proteins containing receptors that detect danger signals located in the extracellular milieu, a transmembrane medium, and an intracellular medium allowing the activation signal transduction) receptors and act as adjuvants. However, the action of these impurities is limited to non-human proteins with a pseudo-vaccination activity. The adjuvants are unable to stimulate an immune response based on a T lymphocyte's response independent of B lymphocytes' tolerance breakdown.

Manufacturing Process and Formulation of the Medicinal Product

The finished product's formulation and conservation conditions are important to maintain the therapeutic protein's biological activity and stability.

Two particularly interesting cases could illustrate how important formulation and conservation conditions are. The first case concerns erythropoietin (EPO). Cases of Pure Red Cell Aplasia (PRCA) after treatment by EPO are known, but rare. Nicole Casedevall [2] has shown the incidence of anti-EPO antibodies' formation, exogenous as well as endogenous, after administration of recombinant human EPO (rHuEPO). These cases have occurred after a changed formulation of the finished product, with human albumin used as a stabiliser replaced by polysorbate 80. Different hypotheses to explain the immune system tolerance breakdown after administration of rHuEPO lead, among other things, to the impurities' extraction from the syringes plunger rod stopper, playing "booster" in presence of EPO [12]. During the analysis of batches called into question, no increase of the level of aggregates or of the level of truncated or degraded EPO has been evidenced. In that case, there must have been several factors having fostered the immune system's tolerance breakdown. In particular, the subcutaneous route of administration may be incriminated, as it is known to produce a pseudo-vaccination effect.

The second case involves the conservation conditions of a freeze-dried formulation of interferon α-2a (rHuINF α-2a) that has been stabilised by human albumin. At room temperature, a partial oxidation of rHuINF α-2a has made easier the formation of aggregates with intact interferon and albumin. These aggregates led to the therapeutic preparation's immunogenicity [13].

These cases illustrate how important the finished product formulation study and the evaluation that has to be done of the possible consequences of a change compared to the initial formulation or its conservation conditions are. It is also particularly important to rigorously analyse the levels of impurities issued from the therapeutic protein's system of production.

Patients and Subsequent Treatment Factors

Patient characteristics, as is the case with their genetic statute and type of disease, are known to influence the response and type of immune responses. It is well-known that patients suffering from severe haemophilia with less than one percent of factor VIII, with time, develop inhibitors to the administration of anti-haemophilic factors of plasmatic origin or derived from recombinant DNA technology. The most plausible explanation resides in the absence of recognition of coagulation factors by the immune system as human proteins of the "self" [5].

In the case of EPO above, with its formulation change, only patients with a chronic renal insufficiency presented an immune system breakdown. Cancer patients treated for their anemia by rHuEPO did not present this secondary effect. This illustrates the conditions promoting an immune tolerance breakdown:

- chronic treatment with repeated doses for months, even years;
- absence of concomitant immunosuppressant treatment;
- route of administration (the subcutaneous injection is more immunogenic than intra-muscular, itself more immunogenic than intra-venous injection).

Case of Monoclonal Antibodies

The first monoclonal antibodies, of mouse origin and obtained as early as 1975, were produced either by ascite' fluid or hybridoma technology. OKT3 used in the kidney cancer of total mouse origin has demonstrated its immunogenicity after it first administration by the standard route of immune system as protein of "non-self." The anti-monoclonal antibodies were neutralising and blocked any repetitive ulterior use.

Technological developments, notably applying the recombinant DNA technology to the production of monoclonal antibodies, has led to the conservation of non-human active variable parts and their graft on the constant humanised parts of immunoglobulin. In that, in this way, have been obtained monoclonal antibodies called chimeric, then humanised, and finally totally human. Today, completely humanised antibodies are obtained through developing technologies such as the phage display, meaning presentation of peptides on the surface of filamentous phages or in transgenic animals.

Humanised monoclonal antibodies are less or very little immunogenic, even if at a very low level there persists a possible induction of antibodies. Treatments by monoclonal antibodies generally require an injection of large and repetitive doses in the range of several hundreds of mg. These doses must be examined with the required level of impurities, notably as far as aggregates are concerned. Even if purification leads to a level lower than 0.5 % of contaminant aggregates, the injected amounts are without commune measure with the other recombinant therapeutic proteins for which the quantity of injected protein matter is in the range of one microgram or one hundred nanograms. Real attention must be paid in that context. It is difficult (based on results of published or conducted studies in the course of therapeutic trials), to closely follow the level of formed antibodies. In fact, the level of circulating and persistent monoclonal antibodies may mask the formation of induced antibodies. Concomitant treatments of patients by cancer medicinal products acting on the immune system or treatments by immunosuppressants may reduce antibodies' neo-formation for these patients. It remains that attention must be paid to the potential immunogenicity of monoclonal antibodies, for they got properties that contribute to it [14]. They can, by themselves, activate T lymphocytes and complement pathways. It has been shown that the loss of glycosyled N-linked chains of the Fc part of immunoglobulin, (as it reduces the activity of the Fc function), may lead to a reduced immunogenicity.

The next authorisation of monoclonal antibody biosimilars will require that their immunogenicity to be particularly checked. This will imply that a similarity demonstration at the quality level be especially well-studied during comparability studies at the level of active substance and finished product. These studies will surely have to be supplemented by more complete safety comparative studies.

Conclusion

Numerous factors influence therapeutic proteins' immunogenicity. At this date it is not possible to completely predict therapeutic protein immunogenicity before the implementation of clinical trials. Immunogenicity is an event that may generally occur with therapeutic proteins. Clinical consequences may vary. The presence of aggregates or the formation of aggregates in the formulation of a therapeutic protein is one of major factors known for increasing immunogenicity. It has been shown that changes in the manufacturing process and in the finished product formulation contribute to modifying the preparation's immunogenicity. If the consequences of these changes are difficult to predict, in vitro tests and tests on immunocompetent transgenic mice are being developed and could lead to an evaluation before clinical trials. These tests do not give any information completely predictive of the therapeutic protein's immunogenicity, but may allow a comparison a formulation to the other, a copy to its reference.

In particular formulation and sources of production of biosimilars, medicinal products should be explored. If regulatorily speaking, the formulation must be identical to the reference product, production cells and purification techniques will always be different. Due to this fact, contaminants coming from the production system will always be different. Thus the biosimilar's immunogenicity in therapeutic condition must be particularly well explored. Most therapeutic proteins induce antibodies in a small number of patients. Post-MA monitoring is especially crucial. Consequently European regulatory authorities ask for a risk management plan to be put in place by pharmaceutical companies once they've got the marketing authorisation.

Publications on immunogenicity cases have shown how important it is to ensure the traceability of batches produced and administered to patients in order to be able to more easily locate the incriminated product, and, if needed, batches in which the events could have taken place. Traceability is not only for biosimilars but for all therapeutic proteins that the patient may receive in case interchangeability may be accepted.

References

1. Dubois M (2008) L'immunogénicité des protéines thérapeutiques dans Développement de techniques analytiques pour l'évaluation des protéines théra-peutiques et des biomarqueurs par spectrométrie de masse, Thèse de Doctorat, Université Pierre et Marie Curie Paris VI. p 106
2. Casadevall N et al (2002) Pure red-cell aplasia and antierythropoietin antibodies in patients treated with recombinant erythropoietin. N Engl J Med 346(7):469–475
3. Gribben JG et al (1990) Development of antibodies to unprotected glycosylation sites on recombinant human GMCSF. Lancet 335(8687):434–437
4. Li J et al (2001) Thrombocytopenia caused by the development of antibodies to thrombopoietin. Blood 98(12):3241–3248
5. Jacquemin MG, Saint-Remy JM (1998) Factor VIII immunogenicity. Haemo-philia 4(4):552–557
6. Chance RE, Root MA, Galloway JA (1976) The immunogenicity of insulin preparations. Acta Endocrinol Suppl (Copenh) 205:185–198
7. Fakharzadeh SS, Kazazian HH Jr (2000) Correlation between factor VIII geno-type and inhibitor development in hemophilia A. Semin Thromb Hemost 26:167–171
8. Shankar Gopi (2007) A risk-based bioanalytical strategy for the assessment of antibody immune responses against biological drugs. Nat Biotechnol 25(5):555–561
9. Schellekens H, Jiskoot W (2007) Immunogenicity of therapeutic proteins. In: Pharmaceutical biotechnology: fundamentals and applications, 3rd edn. Informa Healthcare, New York p 125
10. Crommelin Daan JA (2007) Immunogenicity of therapeutic proteins. In: Handbook of pharmaceutical biotechnology. Wiley, The Netherlands p 816
11. EMEA/CHMP/BMWP/14327/(2006) Guideline on immunogenicity assessment of biotechnology-derived therapeutic proteins

12. Sharma B et al (2004) Technical investigations into the cause of the increased incidence of antibody-mediated pure red cell aplasia associated with Eprex®. Eur J Hosp Pharm 5:86–91
13. Hochuli E (1997) Interferon Immunogenicity: technical evaluation of interfer-on-alpha 2a. J Interferon Cytokine Res 17(Suppl 1):S15–S21
14. Schellekens H (2002) Immunogenicity of therapeutic proteins: clinical implications and future prospects. Clin Ther 24(11):1720–1740

Substitution and Interchangeability

J.-L. Prugnaud

Introduction

The new arrival of the copy of an original medicinal product on the market inevitably poses the questions about the substitution and interchangeability of products. One has to clearly define what the concepts of exchanging a medicine for another are, and who does the exchange, and under what conditions that exchange may be done.

Substitution of Generics and of Biosimilars

In France, substituting an original medicine by a copy is ruled legal for generic medicinal products in the Public Health Code and in the Social Security Code. The substitution is based on the principle that a pharmacist may legally substitute an original medicine whose patent has fallen in the public domain by a generic—copy or the original product—if the latter is listed in generics groups and if the prescribing doctor does not formally oppose in writing on the prescription, to this substitution. Thus the substitution has a regulatory status at a national level. It is the possibility that is given to the pharmacist to replace the medicine corresponding to a brand name by another medicine differently named (but whose active substance, strength and pharmaceutical form are identical), so as to ensure the patient with the same treatment. A generic medicinal product marketing authorisation is given after review of an approval application file made of a complete pharmaceutical part and a clinical part on bioequivalence, showing that

J.-L. Prugnaud (✉)
Prugnaud consulting, 13 ave Jean Aicard, 75011 Paris, France
e-mail: jlprugnaud@gmail.com

J.-L. Prugnaud and J.-H. Trouvin (eds.), *Biosimilars*,
DOI: 10.1007/978-2-8178-0336-4_4, © Springer-Verlag France 2013

the generic product has the same bioavailability as the medicine to which it is compared.

The substitution concept is thus attached, in France, to positive lists of generics. Substitution implies their own conditions of marketing, their inclusion on a list of generics and the prescribing doctor's non-opposability.

What About Biosimilar Medicinal Products?

The definition of a biosimilar medicinal product is very precise (see chapter "From the biosimilar concept to the MA") and states that if it does not fill the conditions of definition of a generic medicine, notably because of "differences linked to raw material or differences between manufacturing processes of the reference biological product," preclinical and clinical trials results must be provided for its approval. Consequently, a biosimilar medicinal product approval application file is different from a generic medicine approval application file. The Code of Public Health refines this approach, stating that biosimilars are not to be included in lists of generics. Stemming from that fact, biosimilar medicines cannot be substituted by the pharmacist. This approach is French and subject to national regulation. Although biosimilars' MA is European and obtained through a centralised procedure, substitution is an approach which varies country by country, according to rules that regulate, among others, medical coverage through social protection.

Interchangeability: Suggested Definition

In the Code of Public Health, it is not said that biosimilar medicinal products are not interchangeable; that is to say that a reference medicine cannot be exchanged for a biosimilar. In fact, nothing legally forbids the interchangeability of an original biological medicinal product by a similar biological medicinal product in compliance with MA's indications. But this exchange falls under a medical act of prescription under the sole responsibility of the attending doctor. From this, a definition of interchangeability may be described as "the possibility, by a medical prescription, to exchange an original medicine for a copy and vice versa". This concept of interchangeability as so defined (that is a "possible exchange") will be better understood in European countries as well as in other regions of the world than the concept of substitution that implicitly or explicitly implies a regulatory overlook according to the concerned country.

Biosimilars' Interchangeability and Conditions to be Implemented

Are biosimilar medicinal products interchangeable and what conditions have to be met for an optimum change?

All biosimilars that have to this date been granted a European marketing authorisation are copies of proteins derived from recombinant DNA technology. Moreover, the marketing authorisation of these biological medicinal products is supplemented in France by particular conditions of prescription and dispensing falling under the statute either of hospital reserve or initial hospital prescription. Some medicinal products reserved for hospital use may however be resold according to outpatient needs by hospital pharmacists. When they are available in the pharmacies, they are dispensed under the pharmacist's responsibility. In all cases and whatever their dispensing statute, they are not substitutable by the pharmacist. The exchange is decided by the doctor by way of his prescription.

In the interchangeability framework, must biosimilar medicinal product prescription be reserved only for doctors defined in the reference medicinal product statute? The answer has been legally given: biosimilar medicinal products follow the same prescription rules as reference products that they copy.

Generics may be prescribed by their common name (INN International Non-proprietary Name), given by WHO (World Health Organisation) to the medicinal product's active substance. Is this possibility transposable to biosimilar medicinal products? As it has been explained in previous chapters, due to the complexity of biological products and the demonstration of similarity-only (and not of strict equivalence between active substance and finished product) differences, if only minimal, may exist at the complex molecular structure level that, in most cases, is not represented by the sole chemical name attributed by the WHO to the active substance. Currently it appears that only the INN is not representative enough of the biological medicinal products (for instance for differences in glycosylation or on other post-translational modifications in direct contact with the producing cell); consequently, the prescription cannot be made by the INN alone. For reasons of prescription accuracy, patient safety and follow-up quality, it is preferable that the prescription be made with the medicinal product's brand name or under a name that makes the production pharmaceutical company identifiable. Suggesting an evolution for biological medicines' INNs is needed; these suggestions will come from adequate WHO studies.

Biosimilars and their reference products are registered in agreement with a centralised European Community procedure. Hence, all Summaries of Product Characteristics (SPC) are the same in all EC countries. Concerning generics, SPCs are identical between generics and reference medicinal products, except when, for example, a clinical indication is still protected by a patent. Could and should it be the same for biosimilars?

As it has been detailed, biosimilar filing is done through results of comparability tests with the reference product, in preclinical and clinical trials. It therefore seems normal that the information based on these comparative studies assessing safety and efficacy be transcribed in the appropriate sections of the SPC. This angle adds another difference with generics. But it is crucial for a quality information representative of the medicine for the prescribing practitioner.

The possibility of interchangeability must be coupled with the patient's rigorous treatment monitoring and notably with the patient's exact treatment. The risk of multiple products taken all along a treatment imposes traceability necessary for the biosimilar as well as for the reference medicine. It could possibly be facilitated by the new European bar code that gives at the same time the medicine's name but also its batch number and its expiry date. Based on this information, it should be easier to follow, batch per batch, the medicinal products administered to the patient. Today this traceability is not required by the authorities, unlike the special case of blood derived medicines. It is strongly desirable in order to more easily know treatments' chronology and, in case of possible undesirable effects, this traceability will give a better grip on these events' history and chronology.

Interchangeability Practices

If substitution by the pharmacist is not possible, what about implementing biosimilars' interchangeability?

A patient's treatment choice is the attending doctor's responsibility. Medical prescription, in all cases, must give with precision the name of the medicine, its strength (notably in case of several strengths), doses and duration of treatment. For a generic, the prescription may be expressed in INN. For a biosimilar, the practitioner must mention with precision the medicine's name or its chosen equivalent because of the pharmacist's forbidden substitution and other compulsory elements in the prescription. The change of medicine is made by the attending doctor according to criteria linked to the particular patient monitored. Depending on the status of dispensing and prescription defined for the medicine marketing, it is possible that the medicine could be changed only by the doctor who had initiated the treatment, and not by the patient referring attending doctor.

Interchangeability management inside hospitals falls under particular conditions. The selection of medicines for the hospital drug formulary is made by a medico-pharmaceutical panel, the committee for medicinal products and sterile medicinal devices (COMEDIMS).

This panel builds, among other things, the policy enabling the choice of all medical products listed in the drug formulary. Biosimilars' arrival onto the competitive market of medicines imposes a policy of selection for products that are not identical to reference products but only similar, that have no generics statute and that have the same indications, either totally or partially. We repeat that for biosimilars the pharmacist has no right of substitution and that the change is under the prescribing doctor's responsibility.

To make treatment' interchangeability possible without patients taking risks, a change policy has to be defined within the medicinal products committee. This policy must rest on the following criteria:

- the committee puts in place medical recommendations to manage the change; they concern interchangeable products, their conditions for prescription, their equivalence, the modalities in which the change can be made, the patient's particulars, patients for whom the change is possible, patients for whom the change is not desirable or doable (population at risk, population possibly not studied in clinical trials, etc.);

- a procedure describes the change's modalities (drug's prescription, dispensing and administration players, distribution chain, specific validations by the doctor or pharmacist if needed, specific collection of monitoring data; notably according to the risk management plan, to which may the biosimilar be submitted, etc.).

Such a policy makes knowing the biosimilar product totally necessary in its pharmaceutical parameters (formulation, strengths, composition) as well as pharmacological and clinical. Not only does the drug's SPC gives this information, but the European Public Assessment Report (EPAR) supplements the scientific information, as it is a document concerning scientific data issued by EMA (European Medicine Agency). Other information is available for COMEDIMS' drafting of recommendations, notably data published in the literature. Initially those will concern the reference product and not its biosimilar.

The choice made by the committee may be based on the following criteria:

- data provided by EPAR, that have shown the similarity;
- SPCs data;
- the disease(s) that the medicine addresses;
- treatment's chronicity, doses, and periodicity;
- routes of administration—criterion relevant to tolerance and immunogenicity;
- pharmaceutical form and possible differences between the biosimilar and original product;
- pediatric data, if necessary;
- existence of a risk management plan and its implementation;
- number of medicines and their pharmaceutical forms;
- evaluation that can be conducted of the potential risks due to interchangeability;
- competitiveness of hospital market;
- availability of the medicine in town pharmacies;
- frequency and duration of tenders;
- price and/or cost for a treatment.

Interchangeability is dealt with "case by case". The committee's selection must take that into account.

May a hospital have on its drug formulary only one type of treatment—the referent medicine or the biosimilar medicine?

This question deserves to be asked for it impacts the policy of allocations of medicines' tenders managed by hospital pharmacists. Taking into account considerations linked to substitution and the special conditions of interchangeability implementation, it is necessary to have a flexible supply of referent medicinal products and their biosimilars in order to fill no substitutable medical prescriptions. All solutions may be considered from the listing on the drug formulary with or without physical inventory, until available at the wholesaler distributor's of references that have not been chosen for the drug formulary and not directly stored in the hospital. Whatever the solution chosen, it takes into account the specificities of that type of drug: non-substitutable biological medicinal products, possibly interchangeable, under the responsibility of the prescribing doctor who has initiated the prescription and compliance with it.

Conclusion

The arrival of biosimilars on the competitive market of biomedicines draws the players involved in the potential use of these products to engage in a reflection on the concept of their substitution and interchangeability. If we cannot stand back, the lack of knowledge and use of these medicinal products may be a temporary factor in a wait-and-see approach of their prescription. It is possible to define, case by case, as for the drafting of development and registration programs aimed at health authorities, recommendations and modalities of use that will frame the dispensation of these products to patients. These will guarantee the safety and good use of biosimilars.

Further Reading

- Directive 2003/63/EC Definition of Biological product (Part I—3.2.1.1.b)
- Directive 2004/27/EC art. 10(4) Biologics: Similar Biological Medicinal Product
- Lekkerkerker F (2009) Are there safety concerns for biosimilars? International Journal of Risk and Safety in Medicine 21 (2009): 47–52
- Revers L, MA, Phil D, Furczon E, HBSc, MBiotech (2010) An introduction to biologics and biosimilars. Part II. Subsequent entry biologics: Biosame or biodifferent ? Canadian Pharmacists Journal 143 (4): 184–91 (July 2010)
- Simoens S (2008) Interchangeability of off-patent medicines: a pharmacoeconomic perspective. Expert Rev. Pharmacoeconomics Outcomes Res 8(6): 519–21
- Substitution des génériques. Loi n° 2007-248 (art. 8) 26 février 2007
- WHO/BS/09.2110 (2009) Guidelines on Evaluation of Similar Biotherapeutic Products

G-CSFs: Onco-Hematologist's Point of View

D. Kamioner

Introduction

Febrile Neutropenia (FN) is associated with an important rate of morbidity and mortality at a high cost for society. FN is still a major threat for patients undergoing cancer chemotherapy, leading to a loss in quality of life. An increase of mortality that can reach 9.5 % appears after hospitalisation for FN. Several risk factors have been identified in order to evaluate the individual FN risk. These factors are patient-linked: age, general health, but also underlying disease (extension, co-morbidity) as well as the chemotherapy protocol used. In order to prevent FN induced by chemotherapy, an antibioprophylaxis and the prescription of Granulocyte colony-stimulating factor (G-CSF) have been somewhat successfully used.

Neutropenia (NP) and FN may lead to a late administration and/or a reduced dose of chemotherapy, hence influence over the disease's evolution.

For Nicole M. Kurderer, the global hospital mortality is 9.5 %; patients without major co-morbidity have a mortality risk of 2.6 %, when a major co-morbidity risk is associated with a 10.3 % risk and anything more than a major co-morbidity is associated with a mortality risk of at least 21.4 %.

The cost to society is high: the average hospital stay is 11.5 days and the average cost is 19 110 dollars per FN episode; patients hospitalised for more than 10 days (35 % of patients) represent 78 % of total cost.

D. Kamioner (✉)
Service de cancérologie et d'hématologie, Hôpital Privé de l'Ouest Parisien,
14 avenue Castiglione del Lago, 78190 Trappes, France
e-mail: dsk.afsos@gmail.com

J.-L. Prugnaud and J.-H. Trouvin (eds.), *Biosimilars*,
DOI: 10.1007/978-2-8178-0336-4_5,

Table 1 Protocols of chemotherapy associated with a risk of NF >20 %

Breast cancer	Bronchial cancer	LMNH
AC/Docetaxel 5–25	ACE 24–57 SCC	DHAP 48
Paclitaxel AC 40	Topotecan 28 SCC	ESHAP 30–64 Doxo/Docetaxel 33–48
Doce/Carbo 26 NSCC	CHOP 21 17–50	
Doxo/paclitaxel 21–32	VP/CDDP 54 NSCC	
TAC 21–24		

NSCC, Non small cell carcinoma; SCC, small cell carcinoma

The major mortality risk factors for hospitalised patients are fungal infection, BG infections, pneumonia and other lung diseases, brain, kidney and liver pathologies (Table 1).

Hemopoiesis takes place at the bone marrow level, notably at the axial skeleton level and long bones. The purpose of using G-CSF is to mobilize bone marrow stem cells and promote precursors proliferation and differentiation.

Overview

Today three products are available in France: filgrastim (Neupogen®), lenograstim (Granocyte®), and pegfilgrastim (Neulasta®), whose mechanisms of action are recognised as follows:

Filgrastim, r-metHuG-CSF

Filgrastim is a human recombinant factor-stimulating granulocyte colonies produced by DNA recombinant technique based on an *Escherichia coli* strain (K12).

It is indicated in the reduction of neutropenia duration and of incidences of febrile neutropenia in patients treated by cytotoxic chemotherapy for a malignant disease (except for chronic myeloid leukaemia and myelodysplastic syndromes), in the reduction of neutropenia duration for patients receiving a myelosuppressant treatment followed by a marrow graft and having an increased risk of severe prolonged neutropenia, and in mobilisation of peripheral stem cells in circulating blood.

Long term administration of filgrastim is indicated for patients, children, or adults alike, suffering from severe congenital, cyclic, or idiopathic neutropenia, with a level of neutrophils $\leq 0.5 \times 10^9$/l and a history of severe or recurrent infections, to increase the neutrophil count and reduce infectious episodes' incidence and duration.

Finally filgrastim is indicated in the treatment of persisting neutropenias (neutrophil count $\leq 1 \times 10^9$/l) in patients infected by HIV at an advanced stage in order to reduce the risk of bacterial infection when the other options aimed at correcting neutropenia are inadequate.

Pharmacokinetics: there is a positive linear correlation between the dose of filgrastim administered by SC or IV injection and the serum concentration. After SC administration at recommended doses, the serum concentrations of filgrastim are maintained above 10 ng/ml for 8–16 h.

Lenograstim, rHu G-CSF

It is produced by a recombinant DNA technique on Chinese hamster ovarian cells and is indicated for reducing the duration of neutropenia in patients (with non-myeloid neoplastie) receiving a myelosuppressive treatment, followed by a bone marrow graft and having an increased risk of severe and prolonged neutropenia, and in the reduction of severe neutropenia duration and complications associated in patients during well-known chemotherapies, known to be associated with a significant incidence of febrile neutropenia and mobilisation of hematopoietic stem cells in peripheral blood. The safe use of lenograstim has not been shown when used with cancer agents with cumulative or predominant platelet lines' (nitrosourea, mitomycin) myelotoxicity. In these situations, using lenograstim could even lead to an increased toxicity, especially for platelets.

Pharmacokinetics: lenograstim is a factor stimulating neutrophil progenitors; as it has been demonstrated by the increased number of CFU-S and CFU-GM in peripheral blood. It notably increases the neutrophil count in peripheral blood, within 24 h after administration. This increased neutrophil level is dose-dependent on between 1 and 10 µg/kg/day. The use of lenograstim in patients who receive a bone marrow transplant or are treated by cytotoxic chemotherapy significantly reduces the neutropenia duration and associated complications.

Pegfilgrastim

Pegfilgrastim, like filgrastim, is produced by recombinant DNA technique from an *Escherichia coli* strain (K12). The filgrastim produced that way and purified from *E.coli, then* goes through a chemical modification in order to introduce PEG (polyethylene glycol) chains on the molecule. These PEG residues are aimed at slowing down filgrastim's degradation and hence at increasing pegfilgrastim's half-life in comparison with filgrastim's.

It is indicated in the reduction of neutropenia's duration and incidence of febrile neutropenia in patients treated by cytotoxic chemotherapy for a malignant syndrome (except for chronic myeloid leukaemia or myelodisplastic syndromes).

Pharmacokinetics: after a single SC administration of pegfilgrastim, the serum concentration peak appears between 16 and 120 h after injection and serum concentration stays stable during the neutropenia period following myelosuppressive chemotherapy. Pegfilgrastim elimination is not a linear function of the dose; pegfilgrastim serum clearance decreases when doses are higher. As the clearance is self-regulated, pegfilgrastim' serum concentration quickly declines as early as the beginning of the restoration of the neutrophil levels.

Limited data show that pegfilgrastim's pharmacokinetic parameters are not modified in patients older than 65.

Biosimilars

The expiration of patents protecting a whole bunch of biological medicinal products these last years has led, in Europe, to approval application and in some cases to marketing of medicines called biosimilars.

If chemical synthesis is the basis of "simple" molecules, biotechnology's benefit resides in the higher capacity of cells to produce complex molecules such as, for instance, human proteins. Therapeutic proteins have three-dimensional structures of high complexity (see chapter "Biosimilars' Characteristics"). Only a precise configuration of these structures leads to an interaction sufficient enough with receptors and therefore, at the end, their biological action. A modest rise in temperature may, for example, make the protein go to another three-dimensionalstate, which in return may lead to a loss of biological function and an increased immunogenicity.

One can therefore understand why clinical studies are necessary to demonstrate the equivalence between biosimilar and reference medicine in therapeutic conditions. Guidelines of EMEA (European Medicines Agency) define general requirements in terms of quality as well as clinical studies; they require approval application data similar to those of the reference medicine. Requirements differ according the concerned protein (EPO [erythropoietin], G-CSF, etc.) and are defined in specific annexes (http://www.emea.europa.eu/ema/pdfs/human/biosimilar for G-CSF).

While a pharmacokinetic bioequivalence proof is sufficient for generics' authorisation, this authorisation is granted to a biosimilar only on the basis of more extended clinical studies demonstrating a therapeutic equivalence with the reference medicinal product.

Sixteen therapeutic areas are concern: haematology (11 %) and cancer (7 %) represent 18 % of therapeutic areas concerned by biosimilars; add infectious diseases (19 %) and these three specialties represent 37 % (see EMEA and AFSSAPS 2005). The market is very wide if France only is considered, it reaches 8.9 % for G-CSF (however almost three times less than EPO).

Prerequisite of EMA Guideline on Similar Biological Medicinal Products Containing Recombinant G-CSF

Pharmacodynamy Study

The number of neutrophils is a pharmacodynamic marker that depends on G-CSF activity. The pharmacodynamic effect of product tests and reference products must be compared in healthy volunteers.

Clinical Efficacy Studies

Severe neutropenia subsequent to cytotoxic chemotherapy prophylaxis in a homogenous group of patients is the recommended clinical model in order to demonstrate the comparability of the test product with the reference product.

Alternatively, models including pharmacodynamic studies on healthy volunteers may be used in order to demonstrate comparability.

Recommendations of EMA (G-CSF Biosimilars Annexes)

A study conducted on healthy volunteers is a model more sensitive for evaluating rG-CSF efficacy than a study in patients undergoing chemotherapy, because healthy volunteers' bone marrow (contrasting with the marrow insufficiency patients') wholly responds to G-CSF treatment.

CHMP (Committee for Medicinal Products for Human Use) has issued positive opinions for filgrastim's biosimilar products for neutropenia's treatment since February 2008: these biosimilar versions are similar to Neupogen®, a reference product.

Besides original G-CSF, biosimilars have recently showed up: Tevagrastim®, Ratiograstim®, Zarzio®, Filgrastim Hospira®, Filgrastim Hexal®, and Filgrastim Mepha® (Suisse). Some are already on the market, others will be soon.

As written above, the arrival of an original medicine's copy on the market inevitably poses the questions of substitution and interchangeability of products among themselves.

Has to be defined and what is implied:

- the concept of exchanging a medicine with another;
- who does the exchange;
- in what conditions that exchange may be performed.

Although the MA of recombinant proteins called biosimilar is European and granted by a centralised procedure, substitution is an approach that varies from one country to another, according to the rules that concern, among others, the covering of health care expenses in the context of social protection.

However, in the Code of Public Health, it is not said that biosimilar medicinal products are not interchangeable. In fact, nothing legally forbids the interchangeability of an original biological medicinal product with a similar biological medicinal product in compliance with MA's indications. But this exchange falls under a medical act of prescription under the sole responsibility of the attending doctor. From that, a definition of interchangeability may be given as "the possibility, by a medical prescription, to exchange an original medicine for a copy and vice versa." (see chapter "Substitution and interchangeability").

Numerous questions are raised for substitution, interchangeability, and even toxicity of biosimilars.

About G-CSF:

- the products currently on the market are biosimilars of figrastim, not lenograstim: in fact these two products' MAs sensibly differ;

- a pegligrastim's biosimilar will soon be on the market;
- filgrastim's potential toxicity is not identical to lenograstim's.

As for any medicine, long-term undesirable effects may occur in the case of systematic substitution and the monitoring will be difficult, if not correctly traced in the patient's file.

Without entering into polemics that followed EPOs' wide use, G-CSF leukemogenic risk must be checked as well as, bone pains, rise of CA 15/3, etc.

Because of some structure variations inherent to the molecule complexity, filgrastim's biosimilars (like all biosimilars) cannot be strictly identical to the original product. The current limited clinical experience requires a special vigilance during the exchange of an innovative product by a biosimilar and their prescriptions' very rigorous follow-up.

A Key Point is the Prescription, in Practice, of Biosimilars: Under What Name, INN (International Nonproprietary Name) or Brand Name?

If for generics the prescription expressed in INN facilitates the pharmacist substitution, it can't be the same for biosimilars, for which the pharmacist substitution is not authorised in France, and in many other European countries. The INN alone is not and has never been sufficient to determine the medicine's prescription and dispensing. Let's remember that the prescription and dispensing of medicines (and naturally of biosimilars) is defined by an adequate work group of the Afssaps (French medicinal products agency). It is this agency that has recommended all provisions in terms of forbidden substitutions for biosimilars. It is also this agency that took the provisions of prescription and dispensing for biosimilars identical to those of the original products (hospital initial prescription, hospital reserve), resale, availability in retail pharmacies, etc.). The recommendations that we may currently give are based not only on precautionary principle, but on a principle of safety and a possible efficient traceability of the medicine at the patient's file level and at the medicine's dispensing by the pharmacist. Today G-CSFs have to be prescribed under their brand names or the INN followed by the producer's name to facilitate the pharmacist's task and to be sure that the prescription be the best possible description of the product that will be administered to the patient.

Is the Same Interchangeability Possible and/or Reasonable for all G-CSFs' Indications?

As mentioned above, the requirements given in the guidelines on the marketing authorisation filing of G-CSF's biosimilars specify that the clinical model chosen for the comparability of biosimilar and reference is the prophylaxis of severe neutropenia induced by cytotoxic chemotherapy in a homogenous group of patients. Because of a same mechanism of action of G-CSFs in the various MA indications of reference products, the biosimilar's producer may ask for all indications' extension, based on the demonstration of clinical comparability in the recommended model. The European Medicines Agency's MAs are not to be put in question and if all indications have been granted, it is because the biosimilar responds to a positive benefit/risk ratio.

However, the observation duration of clinical trials that have been conducted must be watched. Is this observation duration long enough, does it cover all possibilities to observe all undesirable effects and notably the tolerance specific to immunity? The response is partly given by authorities themselves who ask for a risk management plan in postmarketing of G-CSF biosimilars. This plan is destined to cover the insufficient knowledge of these products' tolerance for longer use.

For the prescribing doctor, the attitude is (in the early days of a biosimilar's availability of reserving his prescription of the medicinal products to first-time accessing patients, and then with the publication of results of risk management plans) to widen the medicines' interchangeability. Such an attitude is not overly cautious; it guarantees a better safety for the patient.

Could Situations Occur When Interchangeability Becomes Non Applicable or When, at Least, The Product Could be Changed with Precautions?

As mentioned above, only a longer experience on medicines' tolerance, through a risks management plan, will adequately help dealing with the issue of interchangeability.

The specific situation of multiple changes supposedly induced by nomadic patients has to be taken into consideration. The prescribing doctor must have access to the patient's whole file with the exact prescription that the patient has received, the treatment's duration, the number of chemotherapy cycles, and thus the number of G-CSF cycles performed. Immune system involvement depends, amongst other things, on the treatment's periodicity, its length and how it is repeated.

Clinical trials conducted on biosimilars give information on peripheral stem cell mobilisation in healthy volunteers. Have pharmacology/pharmacodynamics studies been conducted on enough healthy volunteers to provide complete information on the biosimilar's tolerance? If so, using a biosimilar is no riskier than a new medicine.

The same prudent attitude on changing the medicine is to be considered for congenital, severe cyclic, and idiopathic neutropenia indications, because of the long and repetitive treatment. In all cases re-evaluating the long term tolerance based on risks management plan results is a must.

Could an Hospital Pharmacist Impose the Exclusive use of a Biosimilar, in Agreement with the Hospital Tender?

Today it still looks unlikely, due to insufficient experience and a lack of doctor's training in the use of biosimilars. The attitude recommended above has to be debated with the hospital pharmacist and may be validated by the local medicine commission in which doctors and pharmacists discuss the hospital's therapeutic choices.

Conclusion

Due to biotechnologies' development, drug prescription habits change; biosimilars are a new therapeutic approach to which one will have to adapt.

Special precautions for use involve:

- strict compliance with the MA (Marketing Authorisation)
- for the prescribing doctor: a possible change of product for different prescriptions while, however, it is not logical to change medicines during a treatment when there is no proof of therapeutic identity.

Further Reading

1. Aapro M, Crawford J, Kamioner D (2010) Prophylaxis of chemotherapy-induced febrile neutropenia with granulocyte colony-stimulating factors: where are we now? Support Care Cancer 18(5):529–41
2. Aapro MS et al (2006) EORTC Guidelines for the use of granulocyte-colony stimulating factor to reduce the incidence of chemotherapy-induced febrile neutropenia in adult patients with lymphomas and solid tumours. Eur J Cancer
3. Beney J Les Biosimilaires ne sont pas des génériques. Institut Central des Hôpitaux Valaisans, Sion vol. 1 n 6, août 2009
4. Dictionnaire Vidal (2010)
5. EMEA/CHMP/BMWP/31329/2005 Guideline on Similar Medicinal Products containing Recombinant G-CSF
6. Kamioner D (2008) Les indications des facteurs de croissances leucocytaires. Oncolologie, Springer 10(5)
7. M. Kuderer N et al (2006) Cancer 106(10)
8. Möll F (2008) Les biopharmaceutiques et les biosimilaires. Transport, stockage, préparation et utilisation. pharmaJournal 19:5–8
9. NCCN (2010) Practice guidelines in oncology Myeloid growth factors vol. 1
10. R Jared A et al (2006) When the risk of Febrile Neutropenia Is 20 %, prophylactic colony-stimulating factor use Is clinically effective, but Is It cost-effective? JCO 24 (19):2975–2977
11. Repetto L et al. (2003) EORTC cancer in the elderly task force guidelines for the use of colony-stimulating factors in elderly patients with cancer. Cancer 39:2264–72
12. Schellekens H (2009) Biosimilar therapeutics-what do we need to consider? NDT Plus 2(Suppl1):i27–i36 doi:10.1093/ndtplus/sfn177
13. Smith TJ et al (2006) ASCO update of recommendations for the use of white blood cell growth factors: evidence-based clinical practice guideline. JCO 24(19)

The Oncologist's Pointf of View

C. Chouaïd

Introduction

Anemia affects multiple targets, which explains the strong impact of anemia on patients' quality of life (Table 1).

Anemia also impacts patients' survival: in a meta-analysis of 60 clinical trials correlating survival and anemia, it has been observed that the risk relative to death was 65 % higher [IC 95 %: 54–77] (2) in anaemic patients than in non-anaemic patients. In short, anemia linked to cancers appear to be frequent, to have multiple factors, and to have significant consequences. They are a factor in patients' fatigue and create a tumour hypoxia. They possibly could alter treatments' response and their long-term consequences are not well-known. Thanks to existing medicines, they don't have to be undertreated anymore.

In oncology the main drugs that are involved for treating anemia are iron by oral administration, or preferably in injections, erythropoiesis stimulating agents (ESA), and red cells transfusions. Cancer patients' anemia is under-treated. According to ECAS study (European Cancer Anemia Survey) in breast cancer, conducted on 15,000 patients and 1,000 investigators in 24 countries, only 28 % of patients have been treated for their anemia:

- 7 % were receiving an iron supplementation (average dose 11.7 g/l at initiation);
- 12 % an ESA (erythropoietin);
- and 7 % were receiving blood transfusions (average dose 9 g/dl at transfusion time);
- 74 % of patients, with a hemoglobin level of less than 12 g/dl, were receiving no treatment for their anemia.

C. Chouaïd (✉)
Service de pneumologie, Hôpital Saint-Antoine, 184, Rue du Faubourg Saint-Antoine, 75571 Paris, Cedex, France
e-mail: christos.chouaid@sat.aphp.fr

J.-L. Prugnaud and J.-H. Trouvin (eds.), *Biosimilars*,
DOI: 10.1007/978-2-8178-0336-4_6, © Springer-Verlag France 2013

Table 1 Anemia's impact

Central nervous system (CNS)	*Renal function*
Cognitive function	Reduced perfusion
Mood	Water retention
Cardio-vascular system	*Digestive system*
Tachycardia	Irregular transit
Weakness	
Cardio-respiratory system	*Genital tract*
Effort dyspnea	Menstrual disorders
Dyspnea	Lower libido
Cardiac decompensation	Impotence
Skin	*Immune system*
Reduced perfusion	Immunodeficiency
Pallor	
Coldness	

Erythropoiesis Stimulating

Before 1988, blood transfusion was the treatment for mild to severe anemia. Natural human erythropoietin (hEPO) is mainly produced in the kidney. EPO acts directly on progenitor blood cells located in bone marrow to control erythrocytes' proliferation, differentiation and maturation. It is a 165 amino acid glycoprotein, whose biological activity and pharmacokinetics strongly depend on glycosylation. In 1988, epoetin α (Eprex®) was the first ESA to be approved in the treatment of chronic renal insufficiency anemia. In 1993 only it was approved for the treatment of chemotherapy-induced anemia, in Europe. In 1997, epoetin β (Neorecormon®) was approved for epoetin α' same indications. These two ESAs are glycoproteins produced by recombinant DNA technology (rhEPO [recombinant Human Erythropoietin]). They have a primary sequence similar in amino acids to hEPO's. They differ by the number of isoforms representative of their glycosylation profile. The rhEPOs currently on the market have the following main indications:

- anemia of chronic renal insufficiency patients dialysed or not;
- chemotherapy-induced anemia of cancer patients;
- programmed autologous transfusion.

rhEPOs' mechanism of action is the same for all currently approved indications but required doses strongly differ according to indications, with doses much higher in cancer indications.

In 2001 an ESA analogue of rhEPOs was approved, but with a longer duration of action than darbepoetin α (Aranesp®). It is derived from a modified glycosylation with an increased content in sialic acid, hence different pharmacokinetic properties.

Table 2 Benefits and risks of treatments by EPO and blood transfusions

	ESA	Transfusion
Benefits	Hb level improved and maintained Symptoms improved Transfusion needs reduced No need for veinous port of entry No administration in the hospital	Rapid improvement of Hb level and hematocrit Symptoms quickly improved
Risks	Thrombo-embolic risk Small proportion of non responding patients potential survival reduction in cancer patients not treated by chemotherapy Risk of tumoral progression not proved in some cancers	Reaction to transfusion Congestive heart failure Temporarily improved HB level Risk of Iron overload Risk of viral contamination (Hepatitis B [1:250,000]; Hepatitis C [1:2,000,000]; HIV [1:2,000,000]) Development of multiple allo-antibodies

Lifespan is prolonged, making an injection possible every one or every three weeks in compliance with the SPC (Summary of the Product Characteristics).

Since 2007, five copies of epoetin α (Eprex®) have been granted a Marketing Authorisation by the EMA. Filed with biosimilars' dossiers, these rhEPOs have been considered as similar in terms of quality, safety and efficacy to the reference product.

In 2010 has been licensed an epoetin theta (rhEPO-θ) whose structure is related to rhEPOs α and β, with minor glycosylation differences. Their biological activities as well as its duration of action are similar to rhEPO β, but this drug has not been approved under the biosimilar's statute, because of differences linked to glycosylation. It is authorised in the other rhEPO's indications.

Erythropoietins in Anemia Treatments

ESA's value has been demonstrated on the reduction of transfusion needs and on the quality of life. In a number of studies, transfusion need is significantly reduced (about 20 %) compared to the group not receiving ESA. This result has been confirmed for darbepoetin α in a European study conducted on 705 patients with hemoglobin (Hb) lower than 10 g/dl.

A clear improvement in the quality of life parameters, maximum between 11 and 12 g/dl Hb, is shown in the questionnaires follow-up, taking into account different quality of life aspects, notably fatigue.

Although there are no comparative studies between various rhEPOs, it does not seem that there is a difference in efficacy and impact between epoetins (alpha, beta or zeta) or darbepoetin alpha. On the other hand, rhEPOs' biosimilars marketing authorisations give a response in terms of comparison with the reference product to which they are compared.

In anemia treatments, there are benefits and risks (Table 2).

If the benefit/risk ratio of anemia treatments by ESA is positive for the licensing authorities, it has to be noted that a certain number of risks factors exist and deserve to be known. Thus the population to be treated must be defined, based on a few parameters and when a new treatment must be initiated.

When and for Which Patients can rhEPO Treatment may be Prescribed?

Several publications had the impact on survival of treatments by EPO as an objective. A literature analysis identifies eight studies (out of 59) of good quality and dealing with that theme. They involve 3014 patients. Table 3 presents these eight studies. Overall, it seems that there is a risk of high mortality (in the context of anemia with or without chemotherapy and/or radiotherapy) but essentially when the targeted levels to be reached are above 23–14 g/dl.

These analyses' results must be completed by those of meta-analyses taking into account a larger number of patients. They are presented in Table 4, with the number of studies analysed, the number of patients evaluated, the relative risk of death and the confidence interval at 95 % of the relative risk.

In Bohlius et al. meta-analysis, EPOs significantly decrease transfusion needs for patients with a level of Hb $\leq$ 12 g/dl at treatment initiation, and increase the haematological response defined by an increased Hb level by at least 2 g/dl compared to the initial level. There is a positive effect of EPOs on quality of life parameters. However, the risk of thromboembolic complications is higher in patients receiving EPO in comparison with those who do not receive it, as there is a higher risk of high blood pressure in patients treated with EPO.

Definitive conclusions cannot be drawn concerning EPOs' effect on a local tumor response and/or patients' global survival.

In Bennett et al. meta-analysis [4], the risk of thromboembolic complications was increased in patients receiving EPO versus the ones who did not receive it and the mortality risk was significantly increased in patients receiving EPO versus those who did not receive it.

In the new meta-analysis of Bohlius et al. [5], treatments by EPO plus transfusion have been compared to transfusion only.

With an intention to treat analysis done by independent statisticians who have taken into account the meta-analysis' fixed and random effects, overall, EPOs seem to have increased patients' mortality during the studies' active phase and aggravated the patients' global survival (Table 5). However, there is heterogeneity in the analysed studies:

- patients with a low basal level of hematocrit had a higher risk of mortality;
- patients with a history of thromboembolic complications had a lower risk of mortality.

That is not the case if the data stemming from these studies concern only patients undergoing chemotherapy.

Table 3 Effect of an EPO treatment on global survival

Study/type of cancer (n)	Hemoglobin target	Primary endpoint	Results
CHEMOTHERAPY			
1. Leyland-Jones (BEST) Metastatic breast cancer ($n = 939$)	12–14 g/dl	12 month global survival	Decreased global survival: 70 versus 76 %, $p = 0.01$
2. Hedenus (Amgen 161) Lymphoid tumour ($n = 344$)	13–15 g/dl (M) 13–14 g/dl (F)	Proportion of responding patients	Decreased global survival Relative risk = 1.37, $p = 0.04$
3. Prepare Early breast cancer ($n = 733$)	12.5–13 g/dl	Level without relapse and global survival	Level without relapse and global survival decreased global survival tumor progression acceleration deaths: 10 versus 14 %
4. Thomas (GOG-191) Cervical cancer ($n = 114$)	12–14 g/dl	Progression free survival, global survival, locoregional control	Decreased global survival: 75 versus 61 % Survie sans progression diminuée: 65 versus 58 %
RADIOTHERAPY			
5. Henke (ENHANCE) Head and neck cancer ($n = 351$)	5 = 15 g/dl (H) 5 = 14 g/dl (F)	Global survival, locoregional control	Decreased global survival Relative risk = 1.38, $p = 0.02$ Decreased locoregional control Relative risk = 1.69, $p = 0.007$
6. DAHANCA-10 Head and neck Cancer ($n = 522$)	14–15.5 g/dl	Locoregional control	Locoregional control Relative risk = 1.44, $p = 0.03$
WITHOUT RADIOTHERAPY NOR CHEMOTHERAPY			
7. Wright Non small cell lung cancer ($n = 70$)	12–14 g/dl Quality of life		Decreased global survival Relative risk = 1.84, $p = 0.04$
8. Smith (Amgen 103)	12–13 g/dl	Incidence of transfusion	Decreased global survival Relative risk = 1.3, $p = 0.08$

If these meta-analyses analyse reliable data, however, they show limits of interpretation as for survival data:

- profiles and characteristics of patients studied are heterogenous (comorbidity, tumor stages…);

Table 4 Data from meta-analyses

	Number of studies analysed	Number of patients	Death relative risk	95 % CI
Bohlius et al. 2006 [1]	42	8167	1.08	0.99–1.18
Bennett et al. 2008 [2]	51	>13122	1.10	1.01–1.20
Bohlius et al. 2009 [3]	53	13933	1.17	1.06–1.30

Table 5 Meta-analysis, Bohlius et al. 2009

Population	Mortality			Global survival		
	RR	95 % Cl	P	RR	95 % Cl	P
All cancer patients (n = 13933)	1.17	1.06–1.30	0.002	1.06	1.00–1.12	0.05
Studies with chemotherapy (n = 10441)	1.10	0.98–1.24	0.12	1.04	0.97–1.11	0.26

- target haemoglobin levels differ from one study to the other. The target levels were too high because >12 g/dl, going up to 16 g/dl;
- in some studies patients were not receiving any chemotherapy.

Observing in some studies a tumor progression under an EPO treatment could be explained by:

- target levels of Hb > 12 g/dl, not allowing a tumor hypoxia and affecting the disease progression [6];
- re-expression by some tumor types of EPO receptor on the surface of tumor cells. These seem to respond by a proliferation after stimulation by EPO [7];
- EPO off-label use in patients without chemotherapy.

Following another meta-analysis of 12 randomised trials and 2301 patients, if the EPO treatments comply with the criteria for use of EORTC (European organisation of cancer research and treatment), then in that case there is no impact on survival, tumor progression or mortality by thromboembolic events.

Updating the Guidelines (ASCO/ASH)

Results of these various analyses have led learned societies to offer new recommendations for cancer treatments by EPO (Table 6).

Complying with these recommendations must lead to an efficient anemia treatment in cancer patients and so avoid numerous undesirable effects, including thromboembolic complications.

In conclusion, treatment by EPO is an etiological treatment of symptomatic anemia that must be adapted to the clinical situation. The EPO treatment has to be

Table 6 Learned societies recommendations for EPO treatment in cancer patients before and after 2008

	Before February 2008	Since February 2008
Therapeutic indication	Treatment of symptomatic anemia of an adult cancer patient in chemotherapy for a malignant non myeloid disease	
Initial Hb	=s 11 g/dl	=s 10 g/dl
Target values	Not specified	10–12 g/dl
Not over	13 g/dl	12 g/dl
Treatment	Treatment interrupted if Hb > 13 g/dl	Treatment interrupted if Hb > 13 g/dl
Dose adjustment	Reduced dose to maintain Hb at its original level	Reduced dose to ensure that the adequate minimal dose is used to maintain Hb at a level fit to control anemia's symptoms
References	H&N, BEST	H&N, BEST, AoC Meta-analyses

initiated if Hb level is around <10 g/dl and according to clinical state if Hb level is situated between 10 and 11.5 g/dl, with doses calculated on the basis of weight. EPO should be interrupted in case of absence of response in 6–8 weeks or if Hb level increase is lower than 1 g in 1 month and without any impact on the recourse to transfusion. EPO should also be stopped if the response is too rapid (increase of Hb level > 1 g in 2 weeks). Hb level target is 12 g/dl.

Biosimilars

Since 2007, five copies of epoetin α (Table 7) have been granted a marketing authorisation by the European Medicine Agency. Introduced with biosimilars application files, these rhEPOs have a protein structure similar to rhEPO α- (same number of amino acids, same primary and tertiary structure, same number of N and O-glycosylation). All are produced on Chinese hamster ovary (CHO) mammal cells, able to glycosylase the protein structure. Let's remember that this glycosylation is necessary for protein biological activity. Because of the genetic design of production cellular systems specific to each product, there are minor differences at glycosylation levels, which have no incidence upon the pharmacokinetics of each biosimilar comparatively to the reference product (Eprex®). They are all short action duration ESAs. The analytical comparison gives a similar biological activity.

Active substances' INNs (International non-proprietary names) are selected by the WHO and not by European authorities. One of the active substances has a name different from the reference substance. The request for this designation has been made by the producer to ensure a better traceability for the product.

Table 7 Biosimilars authorised in the EU

Biosimilars			References		
Brand name	Active Substance	Producer	Nom	Active Substance	Producer
Binocrit®	Epoetin α	Sandoz	Eprex®	Epoetin α	Jansen-Cilag
Abseamed®	Epoetin α	Medice Arzneimittel Pütter	Eprex®	Epoetin α	Jansen-Cilag
Epoetine Alpha Hexal®	Epoetin α	Hexal AG	Eprex®	Epoetin α	Jansen-Cilag
Retacrit®	Epoetin zeta	Hospira	Eprex®	Epoetin α	Jansen-Cilag
Silapo®	Epoetin zeta	Stada Arzneimittel AG	Eprex®	Epoetin α	Jansen-Cilag

However, there are minor differences in the protein chain glycosylations' minority forms; they have no consequences upon pharmacokinetics nor upon the biological activity of the biosimilar compared to the reference product. For other biosimilars, the manufacturers have chosen to keep the same INN for the active substance; even in the cases of very minor differences in the protein chain glycosylations' minority forms.

Biosimilars have been approved in compliance with the centralised European procedure "of biological medicinal products similar to a reference biological medicinal product." The files filed at the EMEA comprised a complete quality module and shortened preclinical and clinical modules. All studies conducted in terms of quality, safety and efficacy were comparative to the reference product. The benefit/risk ratio has been judged positive by the European authorities and has led to granting a marketing authorisation for the introduced medicines.

Considering that the mechanism of action of erythropoietin was the same in all indications of the reference medicinal product, European authorities have granted to biosimilars all indications of the reference product based on studies conducted on the relevant model of chronic renal insufficiency patients. However, non-comparative tolerance studies at higher doses used for cancer patients have been conducted, especially for thromboembolic events' follow-up in these patients. These studies have shown these biosimilars' equivalent short term tolerance for cancer patients. For cancer, IV and SC administrations have been accepted. Additional studies have been requested for SC route, for chronic renal insufficiency patients, when the studies have not been conducted because of a contra-indication of this route at the time of the study phases.

Benefiting from a centralised approval application, as the reference products, biosimilars have the same Summary of Product Characteristics in the EU 27 countries.

In general, undesirable effects observed during biosimilars' clinical trials have been comparable to those observed with reference EPO. Three consequent risks are generally associated with EPO treatments:

- erythroblastopenia (PRCA Pure red cell aplasia) events;
- vascular thromboembolic events;
- potential risks of tumor progression.

Post-marketing risk management plans have been introduced to follow in cohort studies the incidences in PRCA; especially in patients suffering from renal insufficiency treated for their anemia. Monitoring of thromboembolic events and of tumour progression has been carried out in additional post-marketing pharmacovigilance plans. Risk minimising plans have also been put into place with mention in the SPC of a treatment contraindication by a biosimilar EPO for patients who had developed PRCA in the aftermath of an EPO treatment. In the SPC, attention is drawn on the possibility to develop PRCA when on EPO. As for thromboembolic events, it is mentioned in the SPC that the target Hb level is 12 g/dl. Finally, the potential risk of tumour progression is mentioned in SPC's *ad hoc* sections.

Two PRCA cases have been observed during clinical trials with ESAs with detection of anti-epoetin antibodies. As the investigations of the cause of these PRCAs are ongoing, European authorities consider it important that current EPO treatments' background be monitored and recorded with brand name and/or INN associated with the producer's name. For all ESAs it is also recommended that the information on the product comprises the mention of keeping the recording of patients' treatments.

What About the Possible Substitution and Interchangeability for EPO's Biosimilars?

Biosimilars' substitution and interchangeability are outside European authorities' competence, leaving it up to regional appreciation; that is to say to EU member countries, the rules to put in place. In France the substitution by a pharmacist of a reference product is not possible, for these products are not registrable on the generic drugs repertory. This is the very definition of biosimilars.

However, interchangeability of an original medicinal product by a biosimilar is possible for doctors with regard to their profession's freedom to prescribe. It has to be done within the Marketing Authorisation conditions, and especially the contraindications and precautions for use.

Four ideas may guide the doctor in his approach of biosimilars' prescription and interchangeability:

- the potential immunogenicity of proteins is known, biosimilars, as well as reference products cannot avoid the risk of developing in some patients this undesirable event. With EPOs, the risk of developing antibodies neutralising both exogenous and endogenous erythropoietin is a major risk for the patient. A patient's follow-up is therefore especially important and any decrease in haemoglobin levels during a treatment that has been efficient before must trigger a search for antibodies and guide the decision to interrupt the treatment. It is important to draw attention to the patient's medicine related background, particularly for cancer patients with hepatitis C treated by interferon and ribavirin in whom the PRCA risk is increased;

- targetting 12 g/dl must be imperative for cancer patients treated by EPO, and help controlling thromboembolic events. Speed and figures of recovery of Hb level must be monitored. Recommendations in terms of figures and duration guide the treatment' interruption (see *infra*);
- the prescription may be done only with a brand name or in INN if it is followed by the manufacturer's name. It make the treatment's traceability easier and more secure;
- prescribing a biosimilar for a patient treated for the first time relates to an original drug prescription.

In all cases, with biosimilars arrival on the biomedicines market, traceability for all products is crucial. It is recommended by European authorities and it is mentioned in the information made available for doctors and patients. This gives an opportunity to recall that any change in the medicinal treatment of a patient implies that this treatment has to be explained to the patient and that all information concerning the medicinal product safe use must be provided.

Conclusion

Five biosimilars of EPO have been authorised by the European Union after demonstration of their similarity with an epoetin α (Eprex®) as reference product in all quality, safety and efficiency aspects. Plans of risk management and minimisation have been put in place to supplement routine pharmacovigilance and better inform doctors and patients on use and precautions to take with these products.

Prescribing EPO biosimilars for cancer patients treated for the first time respond to the same level of precaution and of compliance with the MA as for original drugs. Interchangeability is a medical prescription act that must take into account a number of precautions to ensure the patient's safety. Substitution by the pharmacist is not possible. European authorities recommend the traceability of biosimilars' as well as that of original medicines.

References

1. Groopman JE et al (1999) J Natl Cancer Inst 91:1616–1634
2. Caro JJ et al (2001) Cancer 91:2214–2221
3. Bohlius J et al (2006) J Natl Cancer Inst 98(10):708–714
4. Bennett CL et al (2008) JAMA 299(8):914–924
5. Bohlius J et al (2009) Lancet 373(9674):1532–1542
6. Newland, Black (2008) Ann Pharmacother 42:1865–1870
7. Henke et al (2006) J Clin Oncol 24:4708–4713

Biosimilars: Challenges Raised by Biosimilars: Who is Responsible for Cost and Risk Management?

F. Megerlin

Introduction

The advent of "biosimilars" in the European Union has given three types of hopes: hopes of generic producers, eyeing an important market; hopes of payers (insurance companies, hospitals, patients), expecting significant savings; and hopes of doctors and pharmacists, hoping that competition will stimulate research for ever-improving treatments (second generation of biological medicinal products). This chapter gives an overlook on some key points on management in the context of European biosimilars and also sketches the American approach. Biosimilarity is a concept subject to different regulatory definitions internationally and to different demands for comparability. Due to standard differences across the world, the concept of "biosimilar" has hence to be considered rigorously within its proper regulatory context of market approval (MA).

General Information on Cost Management

Because regulatory, technical and financial barriers to market entry are substantial, the EU experience is still limited. To date only fourteen biosimilar products have been approved representing only three active substances (two recombinant growth hormones, five erythropoietins, and seven GCSFs). Three applications have been withdrawn (for biosimilar insulins), and one has been rejected for inadequate similarity to the reference product (interferon alpha-2a). Only large generic manufacturers or biopharmaceutical companies with highly developed bio-production skills and the resources necessary for long-term investments can

F. Megerlin (✉)
LIRAES, Université Paris Descartes, 4 avenue de l'Observatoire, 75006, Paris, France
e-mail: francis.megerlin@parisdescartes.fr

J.-L. Prugnaud and J.-H. Trouvin (eds.), *Biosimilars*,
DOI: 10.1007/978-2-8178-0336-4_7, © Springer-Verlag France 2013

consider entering the EU biosimilar market. After having been authorised for the "single" European market, biological medicinal products and their biosimilars have their modalities of use determined by member states. Reimbursement, pricing, rules on prescription and dispensing, etc. fall in their remits, not in the EU one. Hence each State has its own specificity, as we have seen with substitution (see chapter "Substitution and interchangeability").

Despite these differences, some challenges are common to all EU member states, hospital facilities, physicians, pharmacists and payers. The first challenge is that a biosimilar's purchase price is only one component of its total cost of use.

Savings Linked to the Purchase of Biosimilars

What Percentage of Price Reduction is Due to Competition?

Expiring patents leads to competition between producers and give hope for lower prices. For chemical drugs, price reduction may reach more than 80 % of original product's price. As a result there is a strong price competition as the copy is deemed "identical" to the original product (that facilitates market penetration). For biosimilars in Europe, the average reduction is about 10–30 % of the reference biological medicinal product (for retail price; up to 80 % discount for hospital prices for some products like EPOs, see below).

How to Explain Less Reduced Prices?

Bioproduction explains the difference: it requires a high level of industrial know-how and has irreducible costs, still very high—in contrast with the manufacturing cost of small chemical entities that are easily characterizable and identically reproducible (see chapter "Biologicals's characteristics"). A biosimilar's development costs are about 80–120 million dollars, as opposed to a generic development cost of 0.4–2 million dollars. This gap is also explained by MA procedures: they are much more demanding, therefore more costly than for generics (about 0.8 million dollars) (see foreword). Market penetration is also more costly due to explanations called for by a new concept ("biosimilarity") and by the required precautions. This levelling up explains the limited number of producers qualified to manufacture biosimilars marketed in the EU, contrasting with the myriad of generic producers. To date four companies have been master-applicants (producers), while all others are only co marketing applicants.

Is this Smaller Price Reduction Discouraging?

No, the price reduction compared to the original medicinal product is, for sure, smaller for biosimilars than for generics. But the percentage of reduction applies to a sensibly higher price. For instance, 30 % of price difference for an annual treatment of $50,000 represents a yearly savings of almost $17,000. The potential savings are therefore appealing to buyers (insurance companies, hospitals, patients more or less covered, depending on their country), while the benefit remains significant for the competing producers. Thus the market is also widely animated by price competition, even if price ranges are smaller. Besides, financial constraints to come for many private and public payers are a powerful thinking incentive.

Strategies Played by Competitors

The response varies depending whether that the medicinal product retail prices are fixed or not by public pricing authorities, that they are included or not in health care packages invoiced by hospitals, etc. That depends on national health systems and backgrounds of usage. In general, innovators have a strategy of lowering the prices of their originator products when the market opens to competitors. That sometimes reduces interest for biosimilar producers but also limits the interest of potential buyers such as hospitals: on top of the purchase price, they also bear the cost of use or of first year(s) of use: (see below). Besides, when there is a negotiation between buyer and producer, price difference between a biomedicine and its biosimilar is not necessarily decisive. According to the national systems, the buyer have to consider the risk that the producer of originator products to whom he is not loyal will make commercial conditions less favourable on other products, etc. These anticompetition practices are illegal but they exist. In some countries, the fact that competition is possible leads to tenders, etc. That depends on hospital purchase regulation. This complex management will determine in part the hospital costs of health care providing, and consequently its financial balance.

Why is the Competition for Getting Hospital Markets so Strong?

Depending on the country, biologics may be dispensed in hospitals only, or need an initial hospital prescription, etc. In the first case, hospitals are the only market. In the latter case, hospitals are the key access to outpatient market, which is much more profitable. Retail prices in the EU are generally "listed prices" and non negotiable. Competition for primo prescription is thus fierce; hospital prescriptions often determine subsequent outpatient prescriptions for several reasons: the patients don't like a change in their drugs' brand name and it is hard to explain "biosimilarity". Already acknowledging that some patients are reluctantly using generics, there is a chance for a bigger reluctance facing biosimilars. The patients don't have an absolute necessity to switch if their health coverage neutralizes the costs; the prescribing doctor must monitor the change. Therefore, competitors try to be first on the hospital markets with very low prices for these products (like EPOs), for the real benefit is often to be made in the ensuing outpatient market: (retail prices). But the penetration of hospital market implies for biosimilars to be referenced by competent authorities, to have demonstrated the expected savings to be generated, and puts into question, beyond these products'purchase price, the real cost of their hospital use.

The Issue of Biosimilars' Cost of Using

The cost of using is made of the addition (at purchase price) of costs supported by the buyer and linked to administrative, logistical, and clinical protocol procedures, etc., implemented. Also, as communication strategies of competitors concerning the risks of using are directed towards doctors, communication strategies

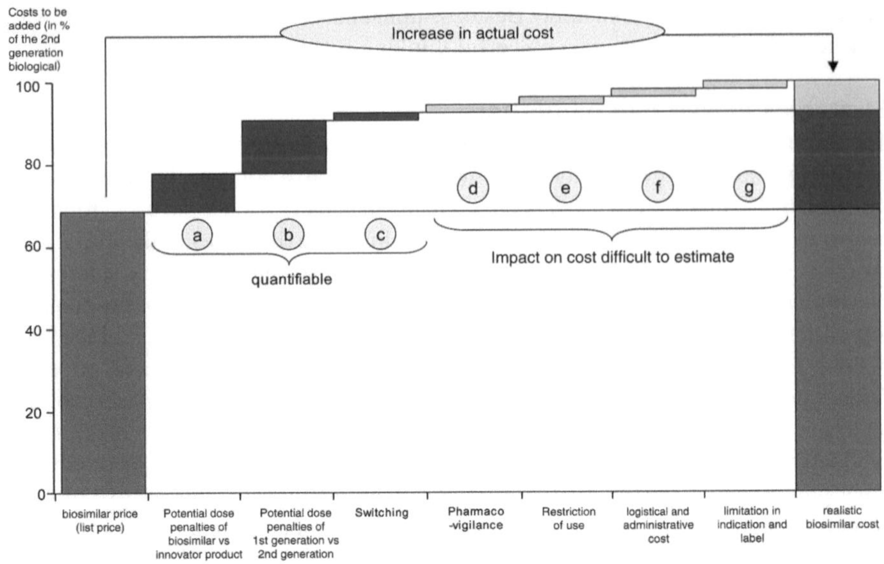

Fig. 1 Theory of cost of using a biosimilar (according to Bols, JEAHP 2008)

concerning the cost of using are directed towards pharmacists when these are purchase-managers.

This pattern in Fig. 1 is emblematic: at the low purchase price of the biosimilar, it adds slices of cost of using. It suggests that the biosimilars' economic benefit is neutralized by these "hidden costs." This pattern was not designed for a particular product. It aims to invite users tempted by a biosimilar to prefer a 2nd generation innovating drug when there is one. However it provides a good basis on which to reflect on a mapping of potential extra costs.

Is There a Non-Equivalence Cost Between Dosages for Biosimilars and Their Reference Biomedicines?

The first slice (a) of additional costs suggests that, to get the same level of effect as the originator product, the first generation biological, a higher amount of substance has to be used, seeming to be a general characteristic of biosimilars. Certainly cases of biosimilars' or biological medicinal products' lower efficacy, even inefficacy, have been reported in international literature; but they were all concerning products marketed *outside* the European Community, probably never assessed in term of their efficacy. A faulty equivalence is not thinkable for biosimilars approved in Europe, given the evaluation criteria of clinical efficacy required at the time of MA submission (see chapter "From the biosimilar concept to the MA"). For biosimilars approved in Europe, no case of a faulty equivalence has been reported to EMA. Consequently, this slice has a documentary interest for medicinal products marketed on foreign markets not or poorly controlled (and in cases of counterfeit reference biomedicinal products), but it is not relevant within the European regulatory framework.

Is There a Non-Equivalence Cost Between Dosages for Biosimilars and Their Second Generation Biomedicines?

The second cost slice (b) suggests that, to get the same effect level as the second generation biomedicine, a larger quantity (or an equivalent-value) of biosimilar should be used. But when that is true (for example, epoietin vs darbopoietin), the same comparison works between the first generation product that was used as a reference for the biosimilar, and the 2nd generation product. In fact, the pattern suggests a preference for 2nd generation biomedicine over all its predecessors. One cannot make the general statement that using 2nd biomedicine is less costly than the use of the 1st generation product: this calls for an analysis per product, and then must be compared the costs of using, not only the biosimilar's cost of using, vs the purchase cost for the 2nd generation biomedicine. Besides, this would first call for a "2nd generation" to be available. This is not systematically the case: for instance, insulin analogs are not in themselves 2nd generation biomedicines; they allow therapeutic "new strategies," but don't replace insulin itself. Such is not the case for erythropoietins alpha or beta which may be replaced by darbopoietin. Ultimately, these 2nd generation products don't necessarily mean that the previous strategies, using the 1st generation products, have to be discarded as the two strategies complete each other.

Is There a Pharmacovigilance Cost Specific to Biosimilars?

The fourth slice (d) additional costs represents pharmacovigilance. This slice is presented as "difficult to quantify". If by pharmacovigilance one means the body of vigilance and notifications applicable in a permanent manner to all medicine, its cost cannot be avoided, be it biomedicine or their biosimilars (for the Eprex® pharmacovigilance case concerned an innovating product, see chapter "Immunogenicity"). If, on the contrary, by pharmacovigilance one means less specifically the user's contribution to the implementation by the biosimilar's MA holder of a "risk management plan", then such a specific cost is not debatable, but is it specific?

Is There a Cost for a Risk Management Plan for Biosimilars?

Yes. The risk management plan is an MA component (see chapter "Immunogenicity"). It is therefore imperative and constitutes a real additional cost (time, clinical and administrative competence). Its significance depends on its follow-up protocol and on the scale of its implementation in the considered hospital or group of hospitals, knowing that the plan is implemented in the whole European Union. But what is important here is that the additional risk management plan cost is not in the realm of biosimilars only: innovating biomedicines must also include a risk management plan in their MA application. The plan is also applied to 2nd generation biomedicines and is a cost of using, etc., that has to be added to their purchase price. Besides, this cost is temporary, as it may disappear all along the risk management plan's evolution and the mastering of initially identified risks. Thus, this slice of additional cost is legitimate but temporary, not specific to biosimilars, and not proportional to a product price.

Is There a Cost if Switching from a Biomedicine to its Biosimilar?

The third slice of cost (c) is supposed to represent a quantifiable cost for switching treatments (see chapter "Substitution and interchangeability"). The possibility of a switch must be considered both ways: from reference biomedicine to biosimilar, and inversely, for instance in case of a break or change in the supply chain. That calls for several reflections. If monitoring a possible switch is a must, switching is not recommended because it would increase the immunogenic risk. It is not a scientific certitude but a highly plausible hypothesis that could be verified only by non ethical trials; the new molecule is not necessarily more immunogenic in itself than its reference, but the switch can unbalance, through consecutive and alternated exposure to molecules similar but not strictly identical, tolerance to the reference molecule that the patient has gradually acquired. Therefore it is not recommended, one way or another. Consequently, the switching cost should not be shown in the graph as systematic and permanent, as suggested by the (c) slice. What is at stake, for producers in competition, is the primo-prescription of their product.

Is There a Cost Linked to a Restricted Use of Biosimilars?

The 5 and 7th cost slices (e and g) generally suggest some limitations in the use, and some restrictions in the authorised indications. What about? A biomolecule may have several treatment indications, several administration modalities (for instance SC or IV for parenteral administration), and a more or less complex pharmacology. Thus, for a biosimilar candidate, when efficacy and/or safety data cannot be extrapolated from a studied indication to a non studied other one, or an administration route to another, additional data are required from the developing producer (see chapter "From the biosimilar's concept to the MA"). For EPO, for example, safety (immunogenicity) data may be extrapolated from SC to IV route, but not in reverse; inversely, when safety and efficacy of an EPO's biosimilar have been demonstrated in patients with chronic renal insufficiency, this biosimilar may be used in other indications. The situation is more complex when several indications are claimed for a single molecule able to interact with several receptors (anti TNF, anti-B cells, anti-VE-VF, etc.). Let's imagine that a producer does not want to bear the cost of additional studies because of the structure of the market he targets, then the use of his biosimilar will be necessarily restricted only to demonstrated indications and modalities, duly justified of regularly extrapolated. This possible restriction must then imperatively be mentionned in the product' SPC as well as the patient information leaflet. In a health care facility, such a restriction may then impose referencing of other "similar" products, whose indications or administration modalities would be, according to the assumption, differentiated. This multiple referencing may induce management costs, or inversely scale savings, depending on clinical needs (more or less specialised), following the eventually regrouped purchase volume, and depending on the prices obtained (the modalities of negotiation or pricing vary with the country). Some biosimilars may be concerned, in agreement with European regulations, by a restricted use, but not in a systematic way. When these restrictions are in place, they do not necessarily induce an extra cost for the health care facility. All depends on the pharmaceutical management of purchase, a discipline that requires a systemic approach, a high scientific expertise and good coordination with doctors. These potential restrictions naturally underline the quality of prescription, dispensing, administration and traceability - as for all medicines.

Is There a Cost Linked to Logistics and Administration of Biosimilars? The sixth slice of costs (f), rightly suggests it. The new references management plus the allocation constraints depending on the product purchase modalities, are real potential costs but vary according to health care structures; it is artificial to represent them in percentage of a product cost that is even not determined. Besides, the same issue exists for second generation biological medicines.

Conclusion

Although it confuses biosimilars approved in and outside the EU, this graph rightly invites a mapping of specific costs that increase product purchase prices. In communication battles, this cost is increased by biomedicine producers, and decreased by biosimilar producers (with both camps exaggerating). In fact, the reasoning cannot settle for all these generalities: it calls for a local and rigourous approach, product by product (both originators and their biosimilars) that would involve all cost of use.

Whatever, risk management is a non negotiable priority. In what terms?

General Information on Risk Management Responsability

By contrast with scarfacing experiences reported in non european-markets (and sometimes confused in the litterature) the European framework of biosimilars authorisation guarantees a high level of quality and safety for these products. To this day, no incident has been reported about authorised biosimilars. What about legal liability? The MA holder's liability leads to technical development without originality: industrial responsibilities respond to the same rules, be the product a biologic or its biosimilar. Let's focus on the responsibility of biosimilar users. This section will deal with the French approach, before dealing with the recent US congresionnal approach.

How the Issue is Dealt with in France

Is There a Specific Legal Responsability for Biosimilars?

No. producers and healthcare providers are exposed to the same legal liability as for other medicines, in case of damages due to a faulty product or a fault. Healthcare providers are essentially liable for malpractice; they are responsible for a faulty product only if the producer is not known. In absence of a defective product or malpractice, national solidarity is activated under "risk development" or "medical hazard," to pay for possible damage. According to french law, a medicine's undesirable effect does not mean that the product is defective, if this potential effect is clearly mentioned with precision in the package leaflet. As a result the producers of biomedicines and of their biosimilars provide thorough information on their leaflets. In fact, warning of a risk on the SPC (Summary of Product Characteristics) only—to which the patient has little or no access—cannot exempt the marketing authorisation holder of his liability. All producers fall under the permanent obligation of pharmacovigilance, health alert, possible batch withdrawal, and leaflet updating:

Is There a Possible Specific Malpractice Associated with Biosimilars?

This question is linked to another one: do health professionals have specific obligations in the context of biosimilars? Except for precautions or restrictions specific to marketing authorisation, the clinical use regulations fall under national competence. In France, there is no specific rule for the use of biosimilars, but biosimilars are not listed in the generics repertoire. Subsequently, their mandatory substitution as for chemical generics is de facto not imposed nor recommeded (see chapter "Substitution and interchangeability"). Moreover, some learned societies have issued recommendations on the use of biosimilars. This type of recommendation affirms a high level of consensus among experts on a state of scientific knowledge. If they are not imposed because of a standard, these recommendations are however source of a latent professionals' obligation. Therefore their content and scope should be studied.

Could Ignoring a Learned Society's Recommendations be a Fault?

The fact that a learned society issues recommendations does not make them a norm. But such recommendations may define means useful or even necessary for ensuring care safety/efficacy, notably, for example, in cases of delay or silence of public sources. Now, besides its clinical application the "state of the art" so objectified may find a legal application: in case of damage that could be avoided or reduced through compliance with the recommendations, a court could consider that a healthcare provider ignoring them had breached a latent "obligation of means," as this professional is supposed to update his scientific knowledge and to do his best for the patient.

Table 1 Summary of learned recommendations

Initial prescription of a biosimilar	Substitution of an innovating product by a biosimilar
	It necessitates a new prescription by the authorised doctor
A biosimilar's traceability must be ensured at injection time	A biosimilar's traceability must be ensured at injection time
An individual prescription follow-up file is put in place and updated by the prescribing doctor	An individual prescription follow-up file is put in place and updated by the prescribing doctor
Complete declaration of observed undesirable effects	Complete declaration of observed undesirable effects
A serum bank is created	A serum bank is created

What is in the Recommendations as far as Biosimilars are Concerned? Again from a methodological point of view, and as an example, here are recommendations issued by Nephrology society, Francophone Dialysis society, and Pediatric Nephrology society (2008). These recommendations are shown in the Table 1.

Let's remember that what interests us here, is, as a procedure optimising the risk management, it is added to the state of scientific and technical knowledge. Although coming from a non-normative source, this procedure becomes virtually effective for practitioners in front of the judge, either civil (private sector care) or administrative (public sector care), hearing the case.

Are These Recommendations Exhaustive and Specific to Biosimilars? No. Recommendations are always in a temporary state of scientific knowledge and concepts. They don't pretend to offer a complete decision algorithm because practicing must integrate all normative constraints: the possible application of a risk management plan for a specified period; obligation to inform patients on the benefit/risk ratio and on the existence (or lack) of treatment alternatives; obligation to collect patients' informed consent on this basis, etc. It should be emphasised that, even if it had not been their intent, these recommendations—and rules—also apply to biomedicines, and not biosimilars only. The risks can perfectly be extrapolated between these two categories of products (potential variability between production sites, even between batches, change or breakdown in the supply chain, immunogenic potential inherent to the product, the treatment duration or the patient's tolerance). This risk management should become reasonably systematic and extend to biomedicines as well as their biosimilars. The same reasoning is valid when setting a serum bank is necessary. Let's remember indeed that reflecting on the risk and its generalisation was born of experience with originator biomedicines, and not with biosimilars (see the Eprex® case). That necessarily leads to the addition of a 3rd column to the table, as follows (Table 2).

Table 2 Extended summary of learned recommendations

Initial prescription of a biosimilar	Substitution of an innovating product by a biosimilar	Initial prescription of a biomedicine
	Substitution of an innovating product by a biosimilar or the reverse! necessitates a new prescription by an authorized doctor	
A biosimilar's traceability must be ensured at injection time	A biosimilar's traceability must be ensured at injection time	A originator biomedicine traceability must be ensured at injection time
An individual prescription follow-up file is put in place and updated by the prescribing doctor	An individual prescription follow-up file is put in place and updated by the prescribing doctor	An individual prescription follow-up file is put in place and updated by the prescribing doctor
Complete declaration of observed undesirable effects	Complete declaration of observed undesirable effects	Complete declaration of observed undesirable effects
A serum bank is created	A serum bank is created	A serum bank is created

Conclusion

In France, legal liabilities are the same for dispensing and prescribing all biomedicines, be they originator or biosimilar. But unbalanced communication on the risk inherent to any biomolecule leads to special focus on biosimilars. It underlines the specific application of follow-up measures, and therefore orientates the attention, even the worry of patients and practitioners alike. Yet, this unbalance in communicating must not lead to ignoring the need of total traceability and vigilance whatever biomedicine (originator as well as its biosimilar) is used. The risk can indeed be induced by possible variability between batches of the same brand, by the way of use, or by the patients profile.

The risks may affect all these products of biological origin, whatever time their brand has been on the market.

Brief Considerations on some international issues

A major issue has been put into light by risk management: the designation of substances of biological origin, and the monitoring of treatments based on products with different batches, production sites or brand names. After a reminder of general reflections, the solution chosen in the US in 2010, will be presented.

The General Issue of the Products' Designation

What is the Substances' International Non-proprietary Name (INN)?

The international non-proprietary name fills in the need for a scientific nomenclature of molecules, that identifies them in an unequivocal and neutral way on an international basis. The scientific name of the molecule is attributed by the World Health Organisation (WHO). It is well distinct from the commercial brand name, chosen by the producer. When the molecule is not protected by a patent anymore, the INN may be used by all producers competing on the market, associated with their own brand name. Using the same INN, affirming the identical scientific classification, is a key element to penetrate markets: it determines the doctors' prescription, and sometimes authorises the "automatic" substitution by the pharmacist (as in the UK). But the problem resides in the fact that bearing the same INN does not mean the same thing for chemical molecules (where molecules can be called identical), and for the field of biological molecules (where the molecules can only be called "similar").

Do Biosimilars Use Their Reference Products' INN?

It is the World Health Organisation (WHO) who attributes INNs. At the European level, the European Medicinal products Agency will only authorise or not the product under the INN filed by the applicant. As a result, as soon as the biosimilarity has been demonstrated according to EU regulation (see chapter "From the biosimilar concept to the MA"), the biosimilar may use the same INN as the reference product, without any obligation to do so. That is only optional for the biosimilars' producer, a strategic marketing decision: some want to put the accent on the biosimilarity, and they use the same INN (for example, several competing EPOs are marketed under the same INN "EPO alpha"). On the opposite, others wish to single out their product and apply for an MA under an INN different from the reference product's (for example a biosimilar uses "EPO zeta" as it INN, when its reference product uses "EPO alpha" as INN). This option is not in contradiction with the biosimilars' European regulation, insofar as the elements constitutive of the "proof of similarity," in the MA biosimilars' regulation context, are not identical to those considered by WHO for allocating an INN, at the request of the considered molecule's holder.

Is There a National Regulation on INN Use?

No. As long as the centralised MA conditions are met and that the authorisation is granted by the European agency, according to the applicant's dossier terms, national authorities are not competent for using that INN or another. But, as an identical INN may in some countries induce a non controlled substitution, some of the Union member states recommend prescribing biomedicines under their brand names rather than under their INNs. That is the case in the UK, on the initiative of the MHRA (Medicines and Healthcare Products Regulatory Agency). In France, the issue is raised in different terms: medicines are generally prescribed under their brand names. Besides and above all, biosimilars are not listed on the repertoire and their substitution, in pharmacies, is de facto not imposed (see chapter "Substitution and interchangeability"). Thus, the fact that a reference product and its biosimilar have the same INN does not make the substitution automatic in France (the substitution is also not "allowed" for chemical molecules that are not listed on the generic repertoire).

Is There a Specific International Regulation of INNs for Biosimilars?

Not today. The question is being discussed at the World Health Organisation level, the only authority competent in INNs matters. Several theses are in competition: producers of originator biomedicines take the position that each medicinal product of biological origin should have its own INN (that would fragment the market and make its penetration harder). On the opposite, producers of biosimilars support the position that their products be allowed to use the same INN as the reference products (that would trivialize them and would make their diffusion on the market easier). The problem is that, in the first hypothesis, attributing specific and different INNs from the reference product to each of its biosimilars (in order to recognize the origin of the molecule) would make the INN system (International Non-proprietary Name) degenerate into an "IPN" system (International Proprietary Name)—the opposite of its neutral scientific classification vocation. In the second hypothesis, the granting of the same INN to a biomedicine and its biosimilars does not respond to the requirements of pharmacovigilance, that needs very precise elements of traceability (see chapter "Substitution and interchangeability"). Other solutions must be found for the classification of biological substances, which are no chemical substances.

Some considerations on the US approach

Regulation of Biosimilars in the US

In 2010, the US have adopted a general regulatory framework for the approval of biosimilars; the guidelines for MA should be developed by the Federal Drug Administration (an American equivalent of the European Medicine Agency). For a large part, the US regulation has adopted the biosimilarity definition and approach elaborated by the European Union. But, while the European regulation has not entered the debate of clinical use, notably as far as substitution is concerned (see chapter "Substitution and interchangeability"), the American regulation have addressed the issue, by taking a position on the qualification of what a "same active substance" means. The denomination of active substances by INN is a fundamental issue in prescription, dispensation, substitution and traceability of products. Yet, as stated above, the context is thorny: although the current solutions for biological substances denomination are not satisfying, the competence of WHO in the classification matter cannot be contested.

How did the US Circumvent the International Issue?

The approach has been very clever: subject to a more precise interpretation, to be given in guidelines that the FDA is bound to issue, the US legislation implicitly subordinates the right, for a biosimilar, to claim its reference product's INN, to the fulfilment of a specific requirement. If the marketing authorisation holder wishes to claim the same INN, he must, above demonstrating the biosimilarity, demonstrate his product's "interchangeability" with the reference product. It does not mean a simple additional quality of the product, but a distinct legal statute (indeed the title of the section is "(k) Licensure of Biological Products as Biosimilar *or Interchangeable*"). The interchangeability proof opens (by inference) the right to the biosimilar's designation by the same INN as the reference product's, the right of substitution by the pharmacist without intervention from the prescribing doctor, and at last the right to a minimal 12 month additional protection of data, for the first biosimilar of a given originator biomedicine whose interchangeability would be demonstrated, (for "rewarding" the developer's effort).

How does US Regulation Define *Interchangeability*?

Interchangeability results from a scientific proof to be assessed by the FDA. According to US 2010 regulation and awaiting for FDA guidelines.

> "(4) SAFETY STANDARDS FOR DETERMINING INTERCHANGEABILITY.
> —Upon review of an application submitted under this subsection or any supplement to such application, the Secretary shall determine the biological product to be interchangeable with the reference product if the Secretary determines that the information submitted in the application (or a supplement to such application) is sufficient to show that — "(A) the biological product — "(i) is biosimilar to the reference product; and "(ii) can be expected to produce the same clinical result as the reference product in any given patient; and "(B) for a biological product that is administered more than once to an individual, the risk in terms of safety or diminished efficacy of alternating or switching between use of the biological product and the reference product is not greater than the risk of using the reference product without such alternation or switch.

This requirement is laudable in its safety intent; but as of today's state of scientific knowledge and available techniques, the demonstration seems to be almost impossible, but for the less complex molecules. That makes few potential applicants, besides insulin for treating diabetes - a market with huge savings potential for the US. Moreover, even if biologics were "interchangeable", US state members can decide wether they authorize the substitution or not.

For the Competing Biosimilars Involved, What are the Consequences of the Lack of *Interchangeability*?

According to US 2010 regulation, the term « interchangeable » or « interchangeability », means that the biological product may be substituted for the reference product without the intervention of the health care provider who prescribed the reference product. Under this section, the "interchangeable product" shall not be considered to have a new active ingredient. In case of non interchangeability, one can infer that the producers of biosimilars will develop a strategy of market penetration equivalent to a new product's. Competition by way of substitution is closed, as it needs medical prescription. Besides, costs and delays required for demonstrating interchangeability make improbable an ulterior benefit, for a biosimilars producer, to change the INN of a product he already commercialises.

That should lead to that, outside of exceptions described above, the reference products and their biosimilars will for long use distinct INNs in the USA. On the other hand, the price competition could be intensified, leaning on a simpler marketing concept.

For American Health Care Professionals, What could be the Consequences of the Lack of *Interchangeability*?

It is doubtful that physicians will take the risk of switching originator biomedicines by "non interchangeable" products, and no insurer will take the risk to push them. Consequently, patients treated before the coming of biosimilars could still be at the innovating product's cost. On the other hand, previously untreated patients (naive) American patients (non previously treated by biologics) will probably be treated at a biosimilar's cost, as "bundled payments" of services and expensive biomedecines could strongly incentivize the choice of biosimilars for first prescription.

Conclusion

When the US regulation makes the substitution improbable, it may, on the other hand, intensify the price competition. Out-of-pocket payment and bundled payments in the US indeed grants to the American prescribing doctor an important social and economical responsibility. Throughout the world, constrained healthcare budgets could fuel the rapid development of biosimilars markets, unless the apparition of new generation of biologics carrying higher value in health convinces payers, potentially the patient himself, and not only the prescribers.

Further Reading

EMEA Workshop on Biosimilar Monoclonal Antibodies, 2 July 2009 – Session 3

Age Biotech Come To (2006) Health Affairs 25:1202–1309

Schellenkens H (2009) Biosimilar therapeutics – what do we need to consider? Nephr. Diag. Trans 2:i27–i36

Pisani J, Bonduelle Y (2008) Opportunities and Barriers in the Biosimilar Mar-ket : Evolution or Revolution for Generics Companies? PWC London

Thompson Pharma's Red Book 2007

Bols T (2008) Biosimilar in Clinical Practice – the Challenges for Hospital Phar-macists. Journal of European Association of Hospital Pharmacists, vol 14(2):33–34

Bols T, Biosimilars in Europe : from approval to clinical practice, Washington University, WDC, sept. 2008

European Generic Medicine Association Handbook for biosimilars, http://www.bogin.nl/files/ega_biosmilarshandbook.pdf

Biosimilars Series: Stakeholder Analysis A Panoramic View of the Emerging Biosimilars Landscape (2008), DATA MONITOR DMHC 2426

Cornes P (2011) The economic pressures for biosimilar drug use in cancer me-dicine. Oncologie, Springer-Verlag. doi:10.1007/S10269-011-2017-9

Megerlin F, Lopert R, Trouvin JH Biosimilars and the European experience: implications for the United States (to be published in Health Affairs)

Afterword

Perspectives: Challenges with Biosimilars

The observant reader of the chapters included in this book will easily have realized that some of the assumptions and rules for biosimilars are, on a case-by-case basis, not easy to prove. For example, how far can one go with the extrapolation of data between different clinical indications? In other words: If for a biosimilar only one key clinical indication has been studied, can one indeed infer efficacy for another indication that is licensed for the reference medicinal product, without any data for the new biosimilar? In the current regulatory approach to this issue, the efficacy and safety of the biosimilar has to be justified or, if necessary, demonstrated separately for each of the claimed indications. This means that there needs to be data, unless otherwise justified. This will most probably not change in future, but many upcoming products that could emerge as biosimilars will require a discussion on one of the central aspects, i.e. whether or not the same mechanisms of action or the same receptor(s) are involved in all indications. This can be difficult to establish for some compounds, for example those which interfere in complex cytokine networks, for diseases where the pathogenesis is not yet clearly defined; or for biological where the mechanism of action is not yet fully understood. Related to this question is the extrapolation of safety: if we accept that efficacy data from one indication sufficiently allow us to include that the biosimilar will work in the other indication as well (as does the reference medicinal product), then can we be sufficiently reassured that also the safety profile would also be similar? Can safety be extrapolated? One could here, on a case-by-case basis, perhaps employ the tool of post-marketing pharma co vigilance and study safety in several indications once the biosimilar is licensed. Another question remains, however: If indeed regulators would accept such extrapolations on scientific grounds, would treating physicians also be willing to accept and use a biosimilar for all indications, especially for more critical clinical indications like anticancer medicines? On scientific grounds this is possible; otherwise a biosimilar would not be licensed in such indication. However, public perception appears to be different at times. This is why for clinicians knowledge of what a biosimilar is and how it is scientifically developed is indeed key for clinicians. Related to this challenge is the question that was already discussed in the opening remarks on this book: If the

J.-L. Prugnaud and J.-H. Trouvin (eds.), *Biosimilars*,
DOI: 10.1007/978-2-8178-0336-4, © Springer-Verlag France 2013

most sensitive model is a particular clinical indication and a particular clinical endpoint, this may not be the most severely affected patient population, or not the most representative clinical presentation for a particular disease. If biosimilarity were established here, would one accept the assumption that the biosimilar will be equally effective (and safe) in more challenging clinical scenarios, e.g. heavily pretreated patients, or patients with more advanced disease? This is also currently under debate, and it is not yet clear what, from a scientific point of view, should prevail: The highest probability of establishing biosimilarity, or the best possible way to establish that the biosimilar also works in clinically challenging scenarios, potentially at the risk that the data is less conclusive since it is significantly confounded by numerous factors.

Could We be too Sensitive?

Another challenge that is indeed interesting is the shift of perception of the power of state-of-the-art physicochemical and biological analytical methods to detect potential differences: Whereas in the past question was "are the methods sensitive enough?", this question has, for some methods, now rather changed to "what do differences mean?", since indeed some methods have reached a high degree of sensitivity, but science is for some aspects not advanced enough to predict or estimate the clinical impact (if any) of measured differences. This will most probably put more burdens on the comparative non-clinical and clinical development. But—what if, for non-clinical studies, in this scenario there is only a relevant animal model (i.e., one in which the biosimilar is pharmacologically active due to the expression of the receptor or an epitope, in the case of monoclonal antibodies) that is a non-human primate species? A powerful powered comparative toxicology study that is designed to detect differences in toxicity may require a large number of animals. This is not only expensive, but also potentially ethically questionable. And, should one focus on on-target toxicity, i.e. toxicity related to pharmacological activity? Here one would indeed have to ask for a relevant animal model. Or, should one study off-target toxicity?

Could one do this in a non-relevant species, for example to test impurities for any undesirable side effects? This would clearly represent a paradigm shift, since data from a non-relevant species is not usually required for biologicals. One would then perhaps, have to focus on the clinical study, and indeed the question then comes back to the extent of extrapolation to other clinical indications one would then be ready to accept, if there were less non-clinical data.

Challenges: Not Always Scientific

Whilst challenging, many points can fortunately be addressed using solid data and sound scientific justifications. However, there are issues that are more related to the perception of a biosimilar—be it justified or not—in the scientific community. How certain perceptions are explained, for example that biosimilars are allegedly "less safe because less well tested", is beyond the author to discuss. Certainly, the reader of this book will now be in a position to appreciate the complexity of the issue, and how high the standards for "true biosimilars" are indeed set. But, there certainly are emerging concerns at least amongst EU regulators about an inappropriate use of the term "biosimilar" and its potential clinical implications. For example, at a recent conference a physician voiced his doubts about biosimilars, since they may not even contain the active substance—here, the term biosimilar was apparently used inappropriately, since this colleague referred to "counterfeit medicines" (i.e., deliberately and fraudulently mislabeled with respect their identity and/or source), not to "biosimilar medicines". The fact that various different terms have emerged internationally for "copy biopharmaceuticals", including "biosimilars" (in the EU), "follow-on biologicals", "subsequent-entry biologicals", "me too biologicals", "biogenerics", "non-innovator" proteins, or "2nd generation proteins" is certainly not helpful. Readers of this book will however surely know what the word "biosimilar" indeed means.

Outlook

Clearly, the further expansion of the biosimilar framework to more complex molecules will raise numerous questions, in fact beyond the question if it is possible to develop more complex biologicals than biosimilars. The question, for some products, will indeed be if it is at all feasible or even financially attractive. For example, complex vaccines, if developed as biosimilars, may be extremely difficult to be compared on a physicochemical and biological analytical level, and a comparative equivalence clinical trial may have to be extremely large to be sufficiently able to detect potential differences to the reference product. The question indeed arises if such a vaccine would be easier to be developed as a "standalone" product, even more so if accepted surrogates for clinical efficacy like immunogenicity parameters for anti-vaccine antibodies exist. With extremely complex medicines like Advanced Therapy Médicinal Products arising (including gène thérapies, cell based médicinal products, and tissue engineered products), one may face complexities that are well beyond the possibilities of biosimilars. But, only time will tell.

Christian K. Schneider
Chairman, CHMP Working Party on Similar Biological (Biosimilar) Médicinal Products Working Party (BMWP),
European Medicines Agency, London, United Kingdom.
Paul-Ehrlich-Institut, Fédéral Agency for Sera and Vaccines, Langen, Germany.
Twincore Centre for Expérimental and Clinical Infection Research, Hannover, Germany.

MIX
Papier aus verantwortungsvollen Quellen
Paper from responsible sources
FSC® C105338
FSC
www.fsc.org

If you have any concerns about our products,
you can contact us on
ProductSafety@springernature.com

In case Publisher is established outside the EU,
the EU authorized representative is:
Springer Nature Customer Service Center GmbH
Europaplatz 3, 69115 Heidelberg, Germany

Printed by Libri Plureos GmbH
in Hamburg, Germany